# Language Acquisition by Children

Edinburgh Advanced Textbooks in Linguistics



TITLES IN THE SERIES INCLUDE

*Essential Programming for Linguistics*
Martin Weisser

*Morphology: From Data to Theories*
Antonio Fábregas and Sergio Scalise

*Language and Logics: An Introduction to the Logical Foundations of Language*
Howard Gregory

*Elements of Formal Semantics: An Introduction to the Mathematical Theory of Meaning in Natural Language*
Yoad Winter

Visit the Edinburgh Advanced Textbooks in Linguistics website at www.edinburghuniversitypress.com/series/edinburgh-advanced-textbooks-in-linguistics.html

# Language Acquisition by Children

## A Linguistic Introduction

Helen Goodluck

EDINBURGH
University Press

Edinburgh University Press is one of the leading university presses in the UK. We publish academic books and journals in our selected subject areas across the humanities and social sciences, combining cutting-edge scholarship with high editorial and production values to produce academic works of lasting importance. For more information visit our website: edinburghuniversitypress.com



Grateful acknowledgement is made to the following sources for permission to reproduce material previously published elsewhere. Every effort has been made to trace the copyright holders, but if any have been inadvertently overlooked, the publisher will be pleased to make the necessary arrangements at the first opportunity.

Chapter 2: Figure 2.1 *Language files*, 2nd edition (The Ohio State University); Figure 2.4 adapted from Eimas et al. (1971: figure 3), *Science*, 171, 303–306. Chapter 3: Table 3.2 adapted from Fortescue (1984/5: 198, table 2), *First Language*, 5, 101–113; Table 3.4 adapted from Clark 2001 (table 13.3); Table 3.7 adapted from Clahsen et al. (1992: table 10), *Cognition*, 45, 225–255. Chapter 4: based on Limber 1973 and Klima and Bellugi 1973. Chapter 5: Figure 5.1 from Valian (2006), *Language Learning and Development*, 2, 251–276; Figures 5.2 and 5.3 from Brooks and Braine (1996), *Cognition*, 60: 3, 235–268; Figure 5.4 from Barner and Snedeker (2005), *Cognition*, 97, 41–66; Table 5.1 abbreviated from Drozd (2002: table 5), *Language Acquisition*, 10, 77–122; Table 5.2 adapted from Noveck (2001: table 2), *Cognition*, 78, 165–185. Chapter 6: Table 6.3 from Yang (2016), *The price of linguistic productivity: How children learn to break the rules of language* (Cambridge, MA: The MIT Press). Chapter 7: Figure 7.2 adapted from Tyler and Marslen Wilson (1981: figure 1), *Journal of Verbal Learning and Verbal Behavior*, 20, 401–416; Figure 7.3 from Trueswell et al. (1999), *Cognition*, 73, 89–134; Table 7.1 from McDaniel et al. (2010), *Journal of Child Language*, 37, 59–94.

Edinburgh University Press Ltd
The Tun – Holyrood Road, 12(2f) Jackson's Entry, Edinburgh EH8 8PJ

Typeset in 10/12 Minion Pro by
Servis Filmsetting Ltd, Stockport, Cheshire

A CIP record for this book is available from the British Library

ISBN 978 1 4744 5815 3 (hardback)
ISBN 978 1 4744 5817 7 (webready PDF)
ISBN 978 1 4744 5816 0 (paperback)
ISBN 978 1 4744 5818 4 (epub)

# CONTENTS


*List of Figures and Tables* viii
*Preface* x

1 Introduction 1
1.1 Knowledge of language: competence and performance 1
1.2 The projection problem 2
1.3 Universal Grammar 2
1.4 The development of the field 3
1.5 Outline 4
Further reading 4
Notes 4

2 The Acquisition of Sound Systems 5
2.1 Speech sounds 5
2.2 Phonetics, phonology and language variation 9
2.3 Speech perception in adults and children 10
2.4 Early speech sounds 15
2.5 Features 16
2.6 Child phonologies 17
2.7 Child systems and phonological theory 24
2.8 Ongoing debates 26
Chapter summary 28
Further reading 29
Questions and exercises 29
Notes 31

3 Morphological Development 32
3.1 Inflectional morphology 32
3.2 Derivational morphology 36
Chapter summary 41
Further reading 42
Questions and exercises 42
Notes 42

4 The Acquisition of Syntax 43
4.1 Syntactic structures and Universal Grammar 43
4.2 The outer course of development 51
4.3 Early syntax 52
4.4 Syntax in the preschool years 55
4.5 Later developments 69
4.6 Syntactic development and the recent theory of grammar 71
4.7 Semantic bootstrapping, prosodic bootstrapping and beyond 73
Chapter summary 75
Further reading 76
Questions and exercises 76
Notes 77

5 Further Aspects of Syntactic and Semantic Development 78
5.1 Tense 78
5.2 Aspect and aktionsart 82
5.3 Negation 84
5.4 Modality 86
5.5 Children's knowledge of quantification 88
5.6 More connections between lexical structures and mental structures 93
Chapter summary 95
Questions and exercises 96
Notes 96

6 Cognition, Environment and Language Learning 98
6.1 Innateness 98
6.2 Input and errors 100
6.3 The role of Universal Grammar in language development 101
6.4 Learnability and acquisition principles 102
6.5 Summary: components of a learning model 103
6.6 Some putative challenges to the model 104
6.7 Continuity vs. maturation 107
6.8 Language development and cognitive development 108
6.9 Errors and error correction 112
Chapter summary 112
Further reading 113
Questions and exercises 113
Notes 114

7 Performance Development 115
7.1 Estimating competence 115
7.2 The adult processing mechanism 116
7.3 Children's processing 119
7.4 Sentence production 125

7.5 Separating immature grammars from immature production/ processing mechanisms 127
Chapter summary 129
Further reading 129
Questions and exercises 129
Notes 131

*Appendix 1:* Methods in child language research 132
*Appendix 2:* Resources for child language research 141
*Bibliography* 142
*Index* 160

# FIGURES AND TABLES

## FIGURES


| | | |
|---|---|---|
| 2.1 | The human vocal tract | 6 |
| 2.2 | Prototypical vowels in English and Swedish | 11 |
| 2.3 | Infant discrimination of vowels sounds in English and Swedish | 12 |
| 2.4 | Mean change in response rate as a function of experimental treatments | 14 |
| 4.1 | The Government-Binding Model | 72 |
| 4.2 | The Minimalist Model | 72 |
| 5.1 | Percentage non-past responses | 81 |
| 5.2 | Images to elicit the reading of *all* (experiment 1, Brooks and Braine 1996) | 88 |
| 5.3 | Images to elicit collective vs. distributive readings (experiment 2, Brooks and Braine 1996) | 90 |
| 5.4 | Images to test mass and count syntax | 95 |
| 6.1 | Components of a learning model | 103 |
| 7.1 | Structure of the adult sentence processing mechanism | 119 |
| 7.2 | Mean reaction times (msecs) in identical monitoring for each prose type | 120 |
| 7.3 | Organisation of visual scenario | 121 |
| A.1 | A typical story for a question response task testing the constraint on extraction from a temporal clause | 135 |
| A.2 | The set-up for a preferential looking experiment | 137 |


## TABLES


| | | |
|---|---|---|
| 2.1 | English vowel sounds | 7 |
| 2.2 | English consonants: place and manner of articulation | 8 |
| 2.3 | Some typical child pronunciation errors in the second and third years | 18 |
| 2.4 | Percentage first syllable stress | 24 |
| 3.1 | Age of mastery of seven morphemes | 33 |
| 3.2 | Single-word utterances of Aqissaq, a two-year-old speaker of Greenlandic Eskimo | 34 |
| 3.3 | One- and two-word utterances of children learning Inuktitut | 35 |
| 3.4 | Inherent vs. temporary states in innovative adjectives | 38 |
| 3.5 | Examples of instrument and locatum denominal verbs | 38 |
| 3.6 | Innovative causatives | 38 |

3.7 Inclusion and omission of plurals in German-speaking children's compounds 40
4.1 Early speech of two children 52
4.2 The emergence of sentence types 53
5.1 Some discourse functions of bare DP ellipsis 86
5.2 Percentage of correct responses to what was in the closed box 88
5.3 Percentage correct responses 89
5.4 Percentage collective interpretations 90
6.1 Complex syntax in a person with Williams syndrome: Crystal (age 15) 109
6.2 Complex syntax in the speech of a person with Down syndrome: Mark (age 32) 110
6.3 Dative errors in the speech of children 112
7.1 Number of filled pauses at utterance onset and other locations 126

# PREFACE

A lot has happened in the field of linguistics and language acquisition since the first edition of this book was published. In the preface to the first edition, I complained about the lack of studies that were well-informed from a linguistic perspective. Since then, there has been research from universities across the world which meets the demand of building a bridge between linguistic theory and language acquisition. And so a new challenge arose: I wanted to write a book suitable for upper-level undergraduates, MA students and students looking around for a doctoral topic, and I wanted the book to cover the broad sweep of core linguist areas (phonology, morphology, syntax and semantics) as well as tackling the contentious issues of the biological origins of our ability to learn language; but I wanted the book not to be too long to use in a one-semester course. The result is this book, which follows the outline of the first book, and includes many of the same examples. One way to meet my goal of limited length would have been to cut out much of the research from before 1990. I believe that this would have been a mistake. The most recent research is not necessarily the best research – although some of it is – and the period between 1960 and 1990 laid important foundations on which subsequent studies were built. A problem was to provide enough detail to help the student get to grips with the issues; the solution I have tried to adopt is to give key references – that is, I hope the reader will follow up the names of the authors I refer to, even if I haven't provided enough information to be crystal clear about the issues. And I know that there are many scholars who deserve attention, but lack of space or ignorance has led me to omit them.

For readers who lack knowledge in one or more areas of linguistics, there are many recent textbooks available: for example, phonology texts: Kennedy (2016) and Hayes (2009); morphology: Lieber (2015); syntax: Adger (2003), Radford (2009) and Carnie (2012); semantics: Kearns (2011). Some students who lack the recommended prerequisites may find things a bit tough going when it comes to certain parts of the text and certain exercises. I would say, don't worry! Just keep going until the land isn't so uphill. There are many recent handbooks in the area of language acquisition. On the grounds that it's better to make one's own summary than to rely on the judgement of others, with one or two exceptions I have not consulted them, but this should not imply that a reader of this book would not benefit from doing so.

I have benefited from the advice and stimulation of faculty at the University of Ottawa (particularly Paul Hirschbühler and Marisa Rivero) and former students with whom I continue to collaborate (particularly Kofi Saah and Danijela Stojanović). The same is true for members of the linguistics department of the University of York, where I moved to in 2004. Thanks to Anne-Michelle Tessier (University of British Columbia), Nino Grillo and George Tsoulas (University of York) for reading sections of the manuscript. The comments of two anonymous reviewers helped shape the book, and the series editors, Peter Ackema and Mits Ota, provided painstaking commentary on the whole manuscript. None of these bears any responsibility for misjudgements or inaccuracies. Bernadette Plunkett kindly let me live in her house while I was finishing the book, and Carrie Singleton and Huw Llewelyn-Jones helped with getting the manuscript into shape. Laura Williamson and Richard Strachan at Edinburgh University Press guided me patiently through the publication process. Jill de Villiers and Tom Roeper deserve general thanks, not least for fine-tuning the question response technique that I have used in my research on question formation. Finally, since the first edition was published, I have lost several friends and advisors from across the years: Patrick Griffiths, Geoff Leech, Peter Schreiber, Rosemary Stevenson and Laurie Stowe. This book would surely have been better if I had had their advice.

If you think that you are too small to make a difference,
you haven't spent a night with a mosquito
(African Proverb; with thanks to Daniel Manyika for supplying
this and for other wise words)

CHAPTER 1

# INTRODUCTION

This chapter begins with a review of some basic issues concerning knowledge of language, followed by a brief tour of the history of the field of language acquisition.

## 1.1 KNOWLEDGE OF LANGUAGE: COMPETENCE AND PERFORMANCE

A basic distinction is made between our linguistic *competence* and our ability to use that competence in everyday speech, i.e. our *performance*. A speaker of English will have no trouble saying that both (1a) and (1b) are grammatical and mean the same thing, and that (2b) is not a grammatical paraphrase of (2a):

1 a Tony threw out the chair
  b Tony threw the chair out
2 a Tony walked out the door
  b *Tony walked the door out[1]

An English speaker can make this judgement even if s/he has never thought about these types of sentences before; but without formal instruction, it is very unlikely that the same speaker will be able to give an accurate account of why it is that s/he finds (2b) ungrammatical. The difference between (1a) and (2a) is that in (1a) *the chair* is the direct object of *threw out*, whereas in (2a) *out the door* is a prepositional phrase modifying the verb *walked*; only in the case of a complex verb such as *throw out*, can *out* be shifted to the end of the sentence.

Speech errors exist – utterances which we as native speakers recognise as deviant from the rules of the language. Such errors may affect different types of knowledge: phonology (the rules determining well-formed pronunciation), morphology (the rules dictating well-formed combinations of elements below the word level), syntax (the rules which determine well-formedness of the strings of words that make up a sentence) and semantics (the rules for assigning meaning to a sentence). Some examples are given in (3):

3 a beef needle (for *beef noodle*, a phonological error)
  b hugger-man (for *man-hugger*, a morphological error)

c You can't figure what that out is (for *You can't figure out what that is*, a syntactic error)
d pass the salt (for *pass the pepper*, a semantic error)

A speaker of English will recognise these utterances as deviant, or unintended (in the case of (3d)). For example, (3c) violates the rule of reordering illustrated by (1a–b). Such reordering must preserve the integrity of the object, yet in (3c) the word *out* is placed *inside* the object noun phrase (*what that is*). (3c) is taken from Garrett (1980: 188).

We can consult the judgements of adult native speakers concerning whether an utterance obeys the rules of their language. No such consultation is available by and large for child speakers, and certainly not for very young children (children aged below three years). And so we have the difficulty of deciding whether a child has made a speech error, or has a non-adult rule of grammar that permits utterances that are disallowed in the mature language. Example (3b) is a child utterance of a type found by Clark et al. (1986).

## 1.2 THE PROJECTION PROBLEM

The child is exposed to spoken language (or sign language for those deaf children who have access to signers). On the basis of the phrases and sentences the child hears, s/he somehow abstracts unconscious knowledge of her/his first language. Children do not receive overt instruction in the rules of their language (see Chapter 6 for further discussion). The task of getting from the necessarily limited language of input (the speech the child hears) to implicit knowledge of the complete adult grammar has been called the *projection problem* (Peters 1972) or the *logical problem of language acquisition* (Baker and McCarthy 1981; Hornstein and Lightfoot 1981).

## 1.3 UNIVERSAL GRAMMAR

Among many linguists (like myself) and psychologists, it is the received opinion that a solution to the projection problem must involve a substantial innate component of linguistic knowledge. Such linguists believe that the gap between the evidence available to the child (the speech s/he hears) and the linguistic system the child ultimately constructs is so great that language acquisition can only be accounted for if we assume that children work with knowledge of principles of grammar. The linguistic system involves rules too abstract and complex to be learned without the aid of innate knowledge. The general idea is that the child is equipped with a set of blueprints that define and limit what a human language can be like. This innate knowledge goes under the name of *Universal Grammar*. Knowledge of Universal Grammar will help the child both by providing a set of candidate analyses for the speech s/he hears and by steering her/him away from any number of possible rule systems that are compatible with the input but simply not found in human languages.

The role of Universal Grammar in language acquisition was influentially laid out and discussed by Chomsky (1965: chapter 1). There Chomsky sketched the distinction between *formal* universals and *substantive* universals. Substantive universals are the 'building blocks' of linguistic rules – the vocabulary in which linguistic rules must be stated. An example is the set of articulatory and/or acoustic specifications that characterise speech sounds (see Chapter 2). Formal universals are restrictions on the types of operations linguistic rules can perform and on the way in which linguistic rules interact. For example, syntactic rules generally pay attention to the hierarchical structure of phrases rather than the linear order of words. We have already seen an example of this in (1a–b). The word *out* can occur either to the right or left of the object phrase *the chair*, but it cannot be interposed between the words that comprise the object – this is due to an obedience to the fact that the object is a unit in the hierarchical structure of the sentence. Any rule which dictated, for example, that a word can be moved from third position in the sentence to fourth or sixth position would fail hopelessly with sentences such as (1a) or *Tony threw out the ugly chair.*

## 1.4 THE DEVELOPMENT OF THE FIELD

The period around the publication of Chomsky (1965) was one of tremendous activity in the field of language acquisition. Psychologists such as Roger Brown and his students conducted breakthrough studies on the development of early child speech, some of which are summarised in Chapter 3. At the same time, Carol Chomsky set in train the study of children's comprehension of syntactic structures (Chomsky 1969), posing problems that are still being explored today. In the 1970s, the world of syntax and language acquisition discovered the notion of c-command (a structural principle defining the height on a hierarchical syntactic structure, see Chapter 4), and this determined many studies of the development of, for example, pronoun interpretation (Lust et al. 1980; Solan 1983; and others). Also in the 1970s, there was the beginning of what is now a crucial component of studies in language development: cross-linguistic comparison (see Bowerman 1973). In the 1980s, computer access to corpora of language acquisition data was initiated (see MacWhinney 2000) and new experimental techniques for the study of child language were developed (for example, Hirsh-Pasek et al. 1987). All of these threads in child language studies have been developed in the 1990s and 2000s, as we will see in the chapters that follow.

So, is the field of language acquisition a thriving and harmonious one at the end of the second decade of the twenty-first century? The answer is yes and no. On the yes side, we will see vigorous debates from a linguistically informed point of view in the chapters that follow. On the no side, there are major contrary trends in child language studies, which I think can be fairly characterised as a conflict between a conviction that a commitment to the idea of Universal Grammar is essential to the study of child language, and a contrasting commitment to doing away with a domain-specific basis for the development of language, or at least the properties characterised in the tradition of grammar represented by Noam Chomsky's work, and those that work with the same broad assumptions. This conflict in approaches is represented in a textbook by Ambridge and Lieven (2011). One purpose of this book is to give the student some background with which to evaluate that text.

## 1.5 OUTLINE

Chapters 2 to 5 deal with the acquisition of phonology, morphology, syntax and semantics. The idea that children's grammatical development is guided by innate knowledge of principles of Universal Grammar is assumed to be correct but is not critical to the points made in those chapters. Linguistic analyses are sketched as a framework for evaluating what children know (and do not know) about the grammar they are learning. Each chapter is more or less independent and they need not be read in their order of appearance (although Chapter 5 relies to some extent on the material in Chapter 4). The sixth chapter takes up more general questions about the nature of innate knowledge and learning mechanisms. The last chapter deals with development in performance mechanisms, which is a growth area in recent research.

## FURTHER READING

Read Chomsky (1965: chapter 1) for a discussion of the distinction between competence and performance and the philosophical background to Chomsky's ideas on innateness. The distinction between competence and performance is further developed in Chomsky (1986). (We sketch more recent thinking in Chapter 4.) Bracken (1983) provides some interesting and very readable commentary on the history of ideas pertinent to innateness and language learning.

## NOTES

1. The convention in linguistics is to use an asterisk (*) to mark ungrammatical sentences.

## CHAPTER 2

# THE ACQUISITION OF SOUND SYSTEMS

This chapter outlines the development of phonetic discrimination by infants and the development of rules of phonology. The basic properties of speech sounds are first summarised, and the early ability to perceive distinctions between vowel and consonantal sounds is described. The chapter then outlines the changes and growth of phonological systems, sketching the child's knowledge of segmental systems, tone and stress. Some unsolved problems include the child's use of consonant harmony.

## 2.1 SPEECH SOUNDS

Words can be analysed as a sequence of discrete units – for example, the word *cat* can be analysed as a sequence of three segments. This section describes some of the major properties of sound segments – properties essential to understanding how segments function in phonology.

*How sounds are made.* The vast majority of speech sounds, and all English sounds, involve pushing air out of the lungs and through the mouth (or nose and mouth). The quality of the sound will depend on the shape of the resonance chambers – the mouth, pharynx and nose (see Figure 2.1) and on whether or not the air stream is obstructed.

The most basic distinction is between vocalic and consonantal sounds. In the articulation of vowels, the air stream is only obstructed at the glottis. Two bands of ligament and elastic tissue, called the vocal cords, obstruct the larynx and the air stream must force its way between the vocal cords when a vowel is articulated, causing them to vibrate. But there is no further obstruction as the air passes through the mouth. By contrast, for most consonants, there may or may not be obstruction at the glottis, but there will be some obstruction of the air passage through the mouth or the pharynx.

*Vowel sounds.* The quality of vowel sounds depends on the position of the body of the tongue – whether the tongue is projected towards the front of the mouth or bunched towards the back, and whether it is held relatively high or relatively low. English has a series of front vowels and a series of back vowels. Table 2.1 lists the main English vowels and gives examples of English words in which they occur. The symbols in Table 2.1 and below are those of the International Phonetic Alphabet; they produce a 'one sound, one symbol' system that eliminates the inconsistencies of English spelling, where in many cases the same letter is used for different sounds and different letters are used for the same sound. The reader can check that the tongue

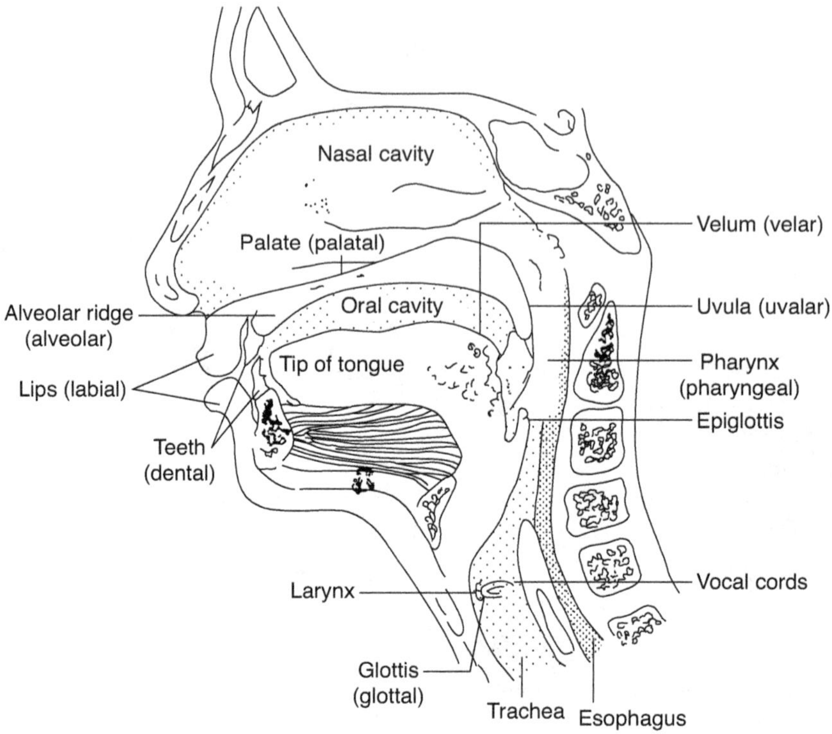


*Figure 2.1 The human vocal tract*

*Source: Language files,* 2nd edition, The Ohio State University.

height positions, etc., are as they are given in the table by saying the example words and comparing the position of the tongue for the different sounds. The mid vowel [ə] (the first and third vowels in the word *banana*) is used in the normal pronunciation of many vowels when they do not bear stress (see section 2.6 below on stress). All of the back vowels in English (except [ɑ]) are pronounced with rounding of the lips, and the front vowels are pronounced without lip-rounding. This type of asymmetry is normal in language, although front round vowels are not uncommon.

*Consonants.* The quality of a consonant sound will depend on the type of obstruction and the place at which the obstruction occurs. Stop sounds involve a very brief complete blockage of air; fricative sounds involve a loose occlusion rather than an absolute blockage. In English, stops are formed by closing the two lips together ([p], [b]), or by touching the tongue to the roof of the mouth at the alveolar ridge ([t], [d]) or the velum ([k], [g]). Fricatives are formed in English by loose contact between the upper teeth and the lower lip ([f], [v]), between the tongue and the upper teeth (the initial sounds in *thin*, [θ], and *then*, [ð]), between the tongue and the alveolar ridge ([s], [z]) and between the tongue and a position slightly to the back of the alveolar ridge (palato-alveolar fricatives, such as the first sound in *sure* [ʃ] and the medial sound in *measure* [ʒ]).

Table 2.1 English vowel sounds

| Tongue height | Tongue projection | | |
|---|---|---|---|
| | Front | Central | Back |
| High | [i] b*ee*t | | [u] b*oo*t |
| | [ɪ] b*i*t | | [ʊ] f*oo*t |
| Mid | [e] b*a*te | [ə] b*a*nan*a* | [o] b*oa*t |
| | [ɛ] b*e*t | | [ɔ] b*ou*ght |
| | | [ʌ] b*u*t | |
| Low | [æ] b*a*t | | [ɑ] b*a*r |

*Notes:*

1. The examples given are based on standard southern British English pronunciation. Details and variants in pronunciation in different varieties of English should not affect the points that follow.
2. The vowels [e] and [o] are in fact diphthongs: combinations of a vowel with a glide sound (see the following section); so too are [i] and [u], although the glide is brief.

All of the above examples come in pairs – [t]/[d], [s]/[z], etc. The difference between the sounds in each pair is not a matter of place of obstruction, or degree of obstruction, but of whether or not the passage of air is additionally interfered with at the glottis. If the vocal cords are close enough together to be set in motion, as they are in the articulation of vowels, then as the air moves through the narrow gap between the cords they will vibrate, and the sound will be *voiced*. If the vocal cords are spread apart, allowing air to pass without obstruction at the glottis, the sound will be *voiceless*. In the pairs of stops and fricatives given above, the first member of the pair is voiceless and the second is voiced; thus [t] is a voiceless alveolar stop and [d] is its voiced equivalent.

In *nasal* consonants there is free passage of the air through the nose. For all sounds described above, the soft part of the back region of the roof of the mouth (the velum) is raised up, so that the entrance to the nasal cavity is blocked and air cannot escape through the nose. If the velum is lowered, the air can pass through the nose and the result is a nasal sound. Each of the stops in English has a nasal variant: [n], the first segment in *night* is the result of a stop articulation at the alveolar ridge plus free passage of air through the nose; [m], the initial sound in *might*, and [ŋ], the last segment in *tang*, are nasals resulting from labial and velar stop articulations, respectively. These nasals are all voiced, as is usually (though not invariantly) the case in languages of the world.

Other types of consonantal sounds include *affricates*, where there is complete closure followed by a gradual, fricative release. English has the palato-alveolar affricates exemplified by the first sounds in *church* ([tʃ], written 'ch') and *judge* ([dʒ]). There are also *liquid* sounds, such as English [l] and [r], formed with semi-free passage of air and somewhat similar acoustically to vowels. *Glide* sounds are also more similar to vowels than to other consonants; like vowels, glides are made without obstruction in the vocal tract. The glides [j] and [w] (the initial sounds in *you* and *witch*) are formed by raising the tongue towards the front and back of the mouth respectively and moving rapidly to or from the position

Table 2.2 English consonants: place and manner of articulation

| | Place | | | | | | |
|---|---|---|---|---|---|---|---|
| Manner | Labial | Labio-dental | Interdental | Alveolar | Palato-alveolar | Palatal | Velar |
| Stop | p, b | | | t, d | | | k, g |
| Fricative | | f, v | θ, ð | s, z | ʃ, ʒ | | |
| Nasal stop | m | | | n | | | ŋ |
| Affricate | | | | | tʃ,dʒ | | |
| Liquid | | | | l | | | |
| | | | | r | | | |
| Glide | w | | | | | j | |

*Notes:*

1. The voiceless member of voiceless/voiced pairs is listed first.
2. Equivalent symbols commonly used in North American transcription are: š = ʃ; č = ʒ; č = tʃ; j = dʒ; y = j.

of the following or preceding vowel; hence the term glide. The back glide [w] is lip-rounded.

Table 2.2 groups the English consonantal sounds described above by place of articulation and manner of obstruction.

*Phonetic features,* A sound segment can be represented as a cluster of properties – a set of plus and minus specifications for *features* that refer to the articulatory or acoustic quality of the segment, or its 'function' in a syllable. The basic distinction between consonant and vowel sounds can be represented in terms of the feature [+/− consonantal]. If we add a second feature [+/− syllabic], reflecting roughly how central in the syllable an element is, we can make a three-way distinction between vowels, consonants and glides, as shown in (1):

| 1 | *vowels* | *consonants* | *glides* |
|---|---|---|---|
| | -consonantal | +consonantal | -consonantal |
| | +syllabic | -syllabic | -syllabic |

Individual segments will be specified for the values of additional features, sufficient to characterise each sound uniquely. The full set of features need not be listed here, and its exact membership is a matter of debate. The features in (2) will be sufficient with respect to the discussion in this chapter.

2 *for vowel sounds*

[+/−high] high vowels are [+high]; mid and low vowels are [-high].

[+/−low] low vowels are [+low]; mid and high vowels are [-low].

[+/−back] back and central vowels are [+back]; front vowels are [-back].

*for consonant sounds*

[+/−continuant] stops are [-continuant]; fricatives are [+continuant].

[+/−anterior] labial sounds (pronounced in the front of the mouth) are [+anterior]; palato-alveolar, palatal and velar sounds are [-anterior].

| | |
|---|---|
| [+/−coronal] | alveolar and palatal sounds are [+coronal] (produced with obstruction by the blade/tip of the tongue); labial and velar sounds are [−coronal]. |
| [+/−voice] | [+voice] sounds are produced with vibration of the vocal cords. |
| [+/−nasal] | [+nasal] sounds are produced with passage of air through the nasal cavity. |

In terms of these features, the word *cat* can be represented as a sequence of segments, with each segment characterised by a set of feature specifications that distinguish that sound from other sounds in English:

| 3 | k | æ | t |
|---|---|---|---|
| | +consonantal | -consonantal | +consonantal |
| | -syllabic | +syllabic | -syllabic |
| | -anterior | -high | +anterior |
| | -coronal | +low | +coronal |
| | -voice | -back | -voice |
| | -nasal | | -nasal |

## 2.2 PHONETICS, PHONOLOGY AND LANGUAGE VARIATION

Phonetics is concerned with the characterisation of speech sounds – how they are produced and perceived and what their acoustic properties are. Phonology is concerned with how sounds are used to distinguish meaning and with regularities that govern the distribution of sounds.

Those phonetic properties of words which are not predictable must be entered in the lexicon (dictionary) of a language; a representation such as that given above for the word *cat* must be part of the lexical entry of that word. Phonetic features which are specified in the lexical entries for a particular language include only the *distinctive features* of the language – features that serve to distinguish between words or morphemes. For example, the feature [+/−voice] is one of the distinctive features of English. If we change the feature specification for voicing in the last segment of *cat*, the difference in sound corresponds to a difference in meaning – *cad* [kæd] does not mean the same as *cat*.

When aspects of pronunciation are predictable, they will not be included in the dictionary entry. There is a difference in the pronunciation of voiceless stops in English. Thus, in word initial position stops are *aspirated* – pronounced with a puff of air on release of the stop. Stops occurring in non-initial position in a cluster of consonants (i.e. after the sound [s] in clusters in words such as *sting* and *stolen*) are unaspirated.

Languages vary with respect to the use to which they put phonetic features. Some features and feature combinations may not be used in a language. Thus, English has no velar fricatives, although velar fricatives are not uncommon in languages. (The voiceless velar fricative does occur in dialects of English, as in Scots pronunciation of the last segment in the word *loch*.) Features may also be distinctive in one language and non-distinctive in another. The feature [+/−aspirate] is an example. Although in English this feature is non-distinctive and its use is predictable, in other languages [+/−aspirate] is used distinctively and the difference between, for example, [k] and [$k^h$] (aspirate [k]) can signal a difference in the meaning of words. Thai and Sesotho (a language spoken in southern Africa) are examples of languages that use aspiration distinctively. Finally, languages may differ with respect to the exact articulatory and acoustic values that are assigned to plus and minus values of phonetic features. In the next section, we will see examples of these types of variation and how infants handle the pertinent contrasts.

## 2.3 SPEECH PERCEPTION IN ADULTS AND CHILDREN

If not all sounds occur in all languages, and those which do occur may or may not be used distinctively, it is plainly part of the task of the language learner to figure out what the repertoire of sounds the language uses is, and which features are distinctive in the language (such as voicing in English) and which are not (such as aspiration in English). The past fifty years have produced remarkable results in the perception of speech sounds by adults and by young infants – babies of less than twelve months. When an adult recognises a sound as, say, the vowel [i] or the consonant [b], what is going on in her/his head? Both in the case of vowels and consonants, speech perception studies have shown that for adults the mapping between the speech signal and the percept involves the imposition of mental categories of an abstract kind on a physical reality that may be quite variable. Moreover, this mapping is either present virtually at birth or develops during the first year of life – and develops in the direction of sensitivity to distinctions that are used in the phonological system of the ambient language. We review here some of the findings for perception of both vowels and consonants.

*Vowels*. The different vowels in and across languages can be characterised in terms of bands of energy, called formants, in the spectrum of the vowels.[1] Kuhl et al. (1992) studied the perception of the English front vowel [i] and the Swedish front vowel [y], which is rounded, unlike its English counterpart. For both vowels, there exists a 'prototype' – a vowel that sounds to the native American English or the Swedish adult ear to be the best exemplar of the vowel. For English [i] the prototype vowel has a first formant at around 350 mels and a second formant at just over 1700 mels.[2] For Swedish [y] the prototype has a first formant of just under 300 mels and a second formant at just under 1600 mels (see Figure 2.2). For adults there is a 'vowel magnet' effect: adults are better at discriminating vowel sounds that differ when they are further away from the prototype. Thus, English speaking adults will be better at discriminating the vowel sounds with a second formant between 1800 mels and 1840 mels than they will be at discriminating between the prototypical vowel with a second formant at 1700 and the vowel with a second formant at 1740 mels, although the absolute difference between the two pairs is the same. Similar effects hold for Swedish-speaking adults for the prototype and non-prototype vowels in their language. It is as if the prototypical vowel draws other vowels near to it, as a magnet would metal pieces.

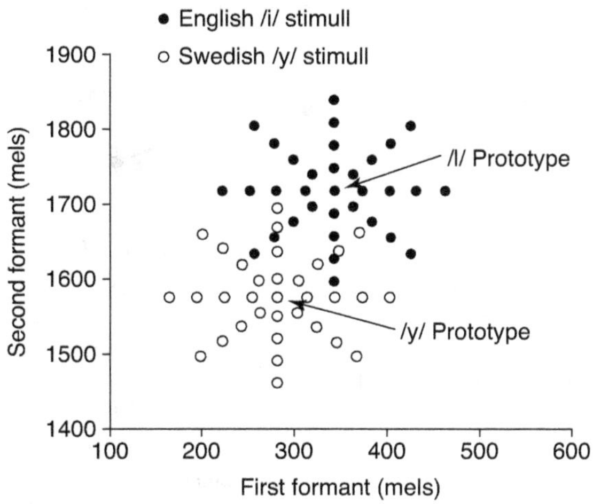


*Figure 2.2 Prototypical vowels in English and Swedish*

*Note:* Six-month-old infants from America and Sweden were tested with two sets of vowel stimuli, American English /i/ and Swedish /y/. Each set included an exceptionally good instance of the vowel (the prototype) and thirty-two variants that formed four rings (eight stimuli each) around the prototype.

*Source:* Kuhl et al. (1992).

When does this perceptual block (the inability to distinguish between sounds close to the magnet) develop? At around eighteen months, when the child has an active vocabulary of fifty or more words, some of them containing the relevant vowels? The answer is no. Vowel magnet effects appropriate to the ambient language have been found for infants aged six months, while the child is half a year away from her/his first words. The presence of vowel magnet effects for [i] and [y] in six-month-old English and Swedish infants was tested by Kuhl et al. (1992). The procedure was conditioned head turn. Each infant was sat on her/his parent's lap and trained to turn their head towards a loudspeaker when the second of two sounds s/he hears is different from the first. Infants of six months can successfully do this. In the test phase of the experiment, the infant heard either the prototype vowel [i] or [y] and one of their variants. The infants were tested in both languages on the perception of prototype and non-prototype vowels, arranged in rings around the prototypical vowel, as shown in Figure 2.3. The percentage of times the infant turned her/his head on encountering a new sound was recorded. English infants failed to discriminate the prototype from another vowel when the vowel was close to the English prototype more often than they did when a member of the pair was more distant from the prototype. The Swedish infants overall showed a greater ability to discriminate the English sounds. When the Swedish vowels were presented, the reverse was true: English infants showed a greater ability to distinguish the vowels of Swedish than the Swedish infants did, with Swedish infants showing a magnet effect for the Swedish sounds.

*Consonants.* Perception of voicing in stop sounds is dependent on the timing

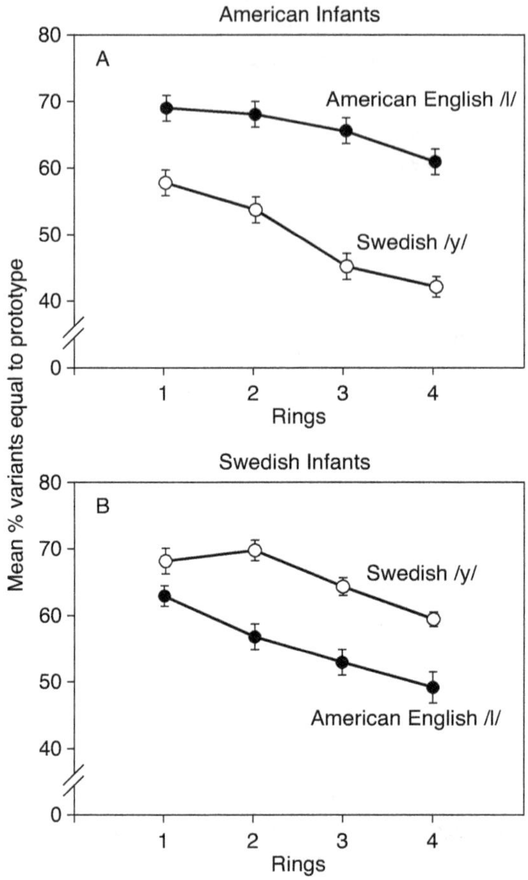


*Figure 2.3 Infant discrimination of vowels sounds in English and Swedish*

*Notes:* Results showing an effect of language experience on young infants' perception of speech. Two groups of six-month-old infants, (A) American and (B) Swedish, were tested with two different vowel prototypes, American English /i/ and Swedish /y/. The mean percentage of trials in which infants equated variants of each of the four rings round the prototype is plotted. Infants from both countries produced a stronger magnet effect (equated variants to the prototype more often) for the native language vowel prototype when compared to the foreign language vowel prototype.

*Source:* Kuhl et al. (1992).

relation between the release of the stop closure and the onset of vibration of the vocal cords.[3] Experiments using synthetic speech stimuli have shown that for adult English speakers a stop sound will be perceived as voiced if the vibration of the vocal cords begins within about thirty milliseconds of the release of the stop closure; if the vibration begins more than thirty milliseconds after the release of the stop closure, the stop will be perceived as voiceless (see Lisker and Abrahamson 1970). The change in perception is quite abrupt. A difference of ten milliseconds of voice onset time

(VOT) in the critical thirty-millisecond VOT region will produce a dramatic change in the perception of sounds as voiced or voiceless. The perception of stops in adults is thus *categorical*: we divide speech events up into discrete categories, such as voiced or voiceless, based on a sharply defined point on the relevant acoustic parameter.

Eimas et al. (1971) performed a clever and ground-breaking experiment, demonstrating that infants of one month and four months are sensitive to the boundary that governs adult perception of voicing in English. Infants can be motivated to suck on a pacifier by an auditory feedback; when an infant sucks with sufficient force, s/he hears a sound. Typically, when the infant catches on to the relation between sucking and feedback, there is a period when the sucking rate increases. Following this, the rate of sucking declines, presumably because the child becomes habituated. In Eimas et al.'s study, the pattern of increased sucking followed by decline was established for each subject with one stimulus type. When the sucking rate for the first stimulus type had declined by 20 per cent or more for two minutes compared with the minute preceding, a second stimulus type was then presented for four minutes. An increase in the sucking rate at the point of changeover can be interpreted as evidence that the infant perceived the difference in the stimuli.

The stimuli in Eimas et al.'s study were synthetically produced syllables consisting of a labial stop plus a low back vowel. Six different stimuli were produced by varying the VOT. Voicing began twenty milliseconds before the stop release, at the stop release, and twenty, forty, sixty and eighty milliseconds after the stop release. For English-speaking adults, stimuli with a VOT of up to and including twenty milliseconds will be perceived as [b]; stimuli with a VOT of forty milliseconds or greater will be perceived as [p]. There were two experimental groups of infants. The first group received stimuli with a VOT of twenty milliseconds after the stop release, followed by stimuli with a VOT of forty milliseconds after the stop release – i.e. the two sets of stimuli straddled the boundary for adult perception of [p] vs. [b]. The second experimental group received two sets of stimuli that did not cross the adult boundary – either stimuli with a VOT of twenty milliseconds preceding the release and stimuli with voicing at the release (both [b] to the adult ear) or stimuli with VOTs of sixty and eighty milliseconds after the release (both [p] to adults). A third control infant group heard the same stimuli throughout (different infants hearing each of the six different stimuli used for the experimental groups). Figure 2.4 displays the mean change in response rate for the two-minute intervals before and after the change in stimuli (or the point at which the change would have occurred, for the control group). The figure shows that a change in stimuli that crossed the adult boundary for [b] vs. [p] produced a marked increase in sucking rate; a change that did not cross the adult boundary produced no such increase (the small rise for one-month-olds was not significant). The control group showed no increase, but a continued decline in sucking rate, as would be expected for the continued presentation of the same stimulus.

Since the work of Eimas et al., a rich literature has been built up concerning the perception of consonant sounds by infants. To give a cursory summary, the young infant is hypothesised to be a 'universal listener' (Werker 1995), able to discriminate all possible distinctions that a language may use. This is supported by many studies which show that infants aged around six to eight months are able to

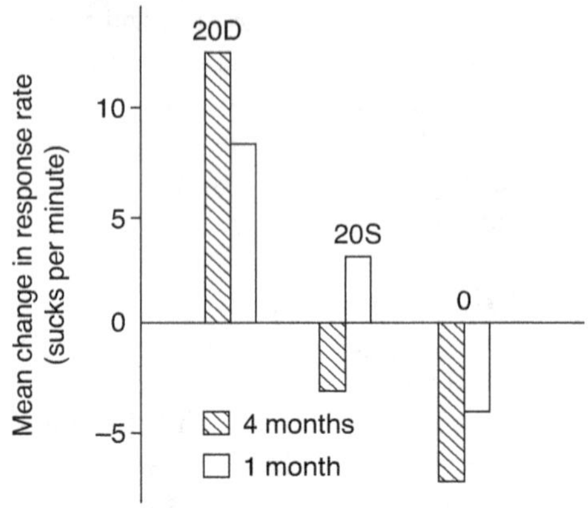


*Figure 2.4 Mean change in response rate as a function of experimental treatments*

*Notes:* 20D = subjects who received stimuli differing by twenty milliseconds VOT across the [p]–[b] boundary. 20S = subjects who received stimuli differing by twenty milliseconds where the difference did not cross the [p]–[b] boundary. 0 = subjects who received the same stimulus throughout.

*Source:* Adapted from Eimas et al. (1971: figure 3).

discriminate sounds that are not used in the ambient language. For example, infants exposed to English can distinguish between the Hindi sounds [Ta] and [ta], the former being a voiceless alveolar stop that is pronounced with the tongue curled backwards, but adult speakers of English cannot make the distinction (Werker et al. 1981). However, there are some limitations on how 'universal' the infant listener is. Lasky et al. (1975) found that four- to six-month-old infants exposed to Spanish, which uses an atypical VOT boundary of 0 milliseconds (as opposed to the English boundary of thirty milliseconds) could discriminate sounds on either side of the English boundary, but not either side of the Spanish boundary. Eilers et al. (1979) found that six- to eight-month-old infants were sensitive to the Spanish boundary, whereas English-learning infants were not. The English boundary is one that many languages use, and thus it may be that at an early stage the child is tuned to the most typical distinctions in languages of the world; for a brief period, sensitivity to the English boundary may be shared by all infants, even if the adult language they are exposed to does not use that boundary. By around eight to ten months, the infant's perception has become adjusted to the boundaries found in the native language. Jusczyk et al. (1993) also demonstrate that infants aged nine months (but not six months) are sensitive to the patterns of stress in English, which uses a basic stressed-unstressed pattern, as opposed to the unstressed-stressed pattern of other languages (see pp. 21–3 below).

The comparison between the results of Kuhl et al. (1992) and those of Werker et al. (1981) reveals a finding that has been echoed in many studies: the language-particular perception of consonants develops somewhat later than for vowels. See Jusczyk (1997: chapters 3 and 4) for a summary of pertinent results.

## 2.4 EARLY SPEECH SOUNDS

Most children begin to produce recognisable words at some point in the second year of life (see Chapter 4). Before that, children pass through a period in which speech-like sounds are produced, with no obvious link to words in the adult language. Playful production of isolated consonant and vowel sounds (typical of four- to six-month-olds) is replaced by *reduplicative babbling*. The child produces a series of consonant-vowel (CV) syllables, in which the individual syllables in each series are identical or very similar to one another. At around ten months, this type of babbling gives way to syllable sequences with more varied members (different consonants and/or vowels) and a wider range of syllable types – VC and CVC in addition to CV (see for example, Stark 1980). The next stage is the production of recognisable words, which may be preceded for some children by a 'silent period'. Vihman (2014: chapter 5) surveys various accounts of stages in early vocalisations, which differ in detail from the sketch just given.

What kinds of sounds do children produce in their babbling and first words? Jakobson (1968, first published in 1941) is a classic study of early speech. Jakobson surveyed diary studies of child speech from many languages and developed an account of the regularities he saw in the data. He writes:

> Whether it is a question of French or Scandinavian children, of English or of Slavic, of Indian or German, or of Estonian, Dutch or Japanese children, every description based on careful observation confirms the striking fact that the relative chronological order of phonological acquisition remains everywhere and at all times the same. (P. 48)

The picture Jakobson sketched can be briefly summarised as follows: during the babbling stage, the child produces a range of speech sounds, some of which may not occur in the ambient language, and some of which may drop out temporarily when the child moves from babbling to first words. For example, velar and palatal stop consonants and affricates may be babbled but then eliminated when the child enters the stage of producing words. In the earliest real-word productions certain vowels and consonants are the first to emerge: [ɑ] and [i] and then [u] or [e] are the first vowels, while the labial stops [m] and [p] followed by the dentals [t] and [n] are early consonants. Fricatives are later acquisitions. Jakobson observed not only that there is a great deal of regularity in the order of emergence in children learning different languages, but the order of emergence mirrors regularities in the distribution of speech sounds in adult languages. The three-vowel system is not merely the most common in early child speech, it is also the most frequent of the two minimal vowel systems used in adult languages (the other being [ɑ], [i] and [e]). The three vowels [ɑ], [i] and [u] are the three most common vowels in languages of the world. Similarly [t] is a very early consonant and it is the 'universal consonant', present in all known adult languages. Jakobson observed that the order of emergence of speech sounds obeyed contingency relations between the occurrence of sounds in adult languages (what he called *laws of irreversible solidarity*): if the presence of sound x presupposes the presence of sound y in the phonological inventories of adult languages, then sound x will not be acquired until sound y has been. The terminology now used is *unmarked* sounds (those that are most frequent in phonological inventories and which are implied by the presence of other sounds) vs. (more) *marked* sounds. Rare sounds can be referred to as the most marked sounds. Notice that frequency in the phonological inventories of the world's languages does not imply frequency of occurrence in the vocabulary of a particular language. For example, the vowel [æ] is a frequent sound in the vocabulary of English, but it is a relatively rare sound in the world's languages. And it is a relatively later addition to children's speech.

Jakobson's account has been criticised on various grounds. Babbling may not be completely free of the influences of the ambient language. Rhythmic properties of babbling and vowel quality in babbling have been shown to differ according to the language the infant is exposed to; so too has the distribution of some consonants (Boysson-Bardies et al. 1992). For many children babbled speech may not be sharply separated from first words, as Jakobson claimed, nor may the inventory of babbled sounds be as distinct from the inventory of sounds in first words as Jakobson implied (Vihman 1992). Nonetheless, Jakobson's basic observation – that there is a strong predictive relationship between the structure of adult phonological inventories and the earliest sounds produced by the child – remains valid. So too does his observation that there is a striking similarity in the babbling of children learning various languages (Locke 1983).

## 2.5 FEATURES

A child may impose patterns on his or her babbling and early speech that are neatly characterised in terms of phonetic features. Gruber (1973) showed that in later babbling, one child in an English-speaking environment preferred sequences in which syllables with alveolar consonants preceded sequences with labials or velars, i.e. sequences such as those in (4) were preferred, and those such as in (5) were avoided:

| | | | | |
|---|---|---|---|---|
| 4 | [də | tə | mə | gə] |
| | alveolar | | labial | velar |
| 5 | *[də | mə | tə | gə] |

Alveolar sounds are produced with an obstruction by the tip or blade of the tongue at the alveolar ridge (the hard ridge directly behind the teeth). The feature [+/−coronal] distinguishes alveolars from labial and velar sounds, where the obstruction is more peripheral (alveolars are [+coronal] and labial and velars are [−coronal]).[4]

Sometimes a child broadens her/his repertoire of sounds that fits exactly with the introduction of a new phonetic feature. For example, when a child begins to make use of the voiced/voiceless distinction, s/he may do so for all the stops s/he uses. (Note that although Eimas et al.'s 1971 study demonstrated that even infants can distinguish voiced from voiceless stops, it is some time before children can use this distinction in producing the language.) When a child begins to use voicing for stops, a three-way distinction [b, d, g] may become a six-way distinction [b – p, d – t, g – k], as we would expect if the feature [+/– voice] is suddenly introduced into the set of features s/he controls. Other children, however, may introduce contrasts piecemeal. For example, Menn (2004, citing her 1976 study) found the child she studied to have the pattern in (6) at one stage,

6 b<br>
t – d<br>
g

Such asymmetries are attested in adult languages and can be described in terms of feature co-occurrence constraints. The asymmetry in (6) can be accounted for in terms of a co-occurrence constraint entailing that [–voice] requires [+coronal]. Levelt and van Oostendorp (2007) develop an account of feature co-occurrence constraints and demonstrate that their system is superior to alternatives. For example, an account that appeals to overall frequency of segments in the input cannot deal with the fact that [t] and [p] are the rarest word initial sounds in child-directed speech in Dutch (the language studied by Levelt and van Oostendorp) and yet they are among the first sounds acquired.

## 2.6 CHILD PHONOLOGIES

Infants recognise the difference between speech sounds, as shown in section 2.3. But this does not imply that they have a phonological system, a system that maps between words in the lexicon and what is pronounced. Other animals display categorical perception (Kuhl and Miller 1975), indicating that the roots of perception may in part be physiological. Only the human child moves to map sounds to words using a full-blown phonological system.

*The representation of first words.* When the child is a little over one year, s/he produces words that are recognisable to the adult. Swingley and Aslin (2002) demonstrated that 14- to 15-month-old infants recognise mispronunciations (e.g. *pity* for *kitty*), suggesting that they have a fairly detailed representation of the phonetics of words.[5]

*Errors of production in child speech.* In what follows we will assume both for adults and for children that there are base forms that are manipulated or evaluated by the phonology. The base form for adults is the lexical entry. The base form for children is frequently taken to be the adult's utterance (see section 2.8 below for some further discussion).

Although the child produces words that are recognisable to adults, nonetheless children do use words that are very often incomprehensible to their parents, even when trained as linguists! Some typical child pronunciations are listed in Table 2.3.

Detailed study of children's pronunciation errors has revealed their systematic nature. Ground-breaking in this regard was Smith's (1973) study of his son Amahl. Amahl produced [æŋu] for the target word *handle*, and [ɛbi:] for the word *empty*. Smith proposed a set of rules from which the child's output could be derived; the following rules were responsible for Amahl's utterances between the ages of 2;2 (years;months) and 2;4 (the rule numbers are taken from Smith):

Table 2.3 Some typical child pronunciation errors in the second and third years, based on Ingram (1986)

| | Examples | |
|---|---|---|
| | Adult word | Child pronunciation |
| *Substitution processes* | | |
| (replacement of one sound by another sound) | | |
| Stopping (a fricative is replaced with a stop) | see | ti: |
| Fronting (the place of articulation is fronted with velar and palatal consonants being replaced by alveolars) | goat | dut |
| Gliding ([w] or [j] is substituted for a liquid) | leg | jek |
| | ready | wedi |
| *Assimilation processes* | | |
| (a sound becomes more similar to an adjacent sound) | | |
| Voicing (consonants tend to be voiced preceding a vowel and devoiced at the end of a syllable) | paper | be:bə |
| | pig | bik |
| Consonant harmony (consonants tend to assimilate in words with the structure $C_1$ V $C_2$ (X)) | duck | gʌk |
| | tickle | gigu |
| | tub | bʌb |
| Progressive vowel assimilation (an unstressed vowel will assimilate to a preceding vowel) | bacon | 'bu:du |
| | flower | 'fa:wa |
| *Syllable structure processes* | | |
| Cluster reduction | play | pe |
| | train | ten |
| | dress | dɛs |
| Final consonant deletion (CVC is reduced to CV) | bib | bi |
| | more | mʌ |
| Unstressed syllable deletion | ba'nana | 'næna |
| | po'tato | 'dedo |
| Reduplication (in a multisyllabic word, the initial CV is repeated) | TV | didi |
| | water | wawa |

*Notes:*
: = length mark
' = stress mark (indicates the following syllable is stressed).

7 Rule 1: If a nasal is followed by a voiceless consonant, the nasal is deleted.
Rule 2: If a nasal is followed by a voiced consonant, the consonant is deleted.
Rule 3: A coronal stop is velarised before /(ə)l/.[6]
Rule 4: An unstressed vowel is raised and backed before /l/.
Rule 6: /l/ is deleted at the end of a word.
Rule 13: /h/ is deleted.
Rule 21: An alveolar consonant is deleted after another consonant.
Rule 25: All voiceless segments are voiced.

These rules can be used to account for the child's pronunciation of the words *handle* and *empty*, as shown in (8):

| 8 | Adult form: | *handle* | [hændəl] | |
|---|---|---|---|---|
| | Child rules: | Rule 3 | / hæŋəl / | |
| | | Rule 4 | / hæŋul / | |
| | | Rule 6 | / hæŋu / | |
| | | Rule 13 | [ æŋu ] | = child pronunciation |
| | Adult form: | *empty* | [ɛmpti:] | |
| | Child rules: | Rule 1 | / ɛpti: / | |
| | | Rule 21 | / ɛpi: / | |
| | | Rule 23 | [ ɛbi: ] | = child pronunciation |

Amahl's rules applied quite generally, not just to the individual words analysed in (8).

Braine (1976) takes issue with some points in Smith's analysis. For example, Rule 1 (which deletes a nasal before a voiceless consonant) and Rule 2 (which deletes a voiced consonant after a nasal) may alternatively be accounted for in terms of perceptual error on the child's part. Vowels in English are lengthened before a voiced consonant. Acoustically, vowels carry cues for the perception of a following nasal, and the lengthening of the vowel as a consequence of a following voiced consonant may thus cause a nasal in a V-nasal-C sequence to be perceived, where it would not be perceived if the consonant were voiceless and the vowel not lengthened. Thus, in part the rules Smith proposes may be an effect of a perceptual filter.

Such reservations aside, it is clear that Amahl has a systematic way of adjusting the adult form of words. The very fact that the adult input can be manipulated in the manner in (8) speaks to the child having in his or her head something like the adult pronunciation. Moreover, children can recognise their own mispronunciation as deviant (adult: 'Did you say "wellow"?'; child: 'No, I said "wellow"'). Thus, the child can be argued in effect to have her/his own phonological system, which takes an input form and converts it to the form that s/he utters. The child's rules very often have a parallel in adult languages, if not the language s/he is learning. For example, children not uncommonly de-voice segments in word final position, pronouncing *bag* as [bæk]; such a de-voicing rule is found in languages such as German.

*Tone.* In tone languages, pitch is used to distinguish words and morphemes. Thus, pitch is a property of words and must be entered in the lexicon of the language (the mental dictionary). For example, Mandarin Chinese has four distinctive tones: high-level, rising, falling and dipping. These terms reflect the pitch levels associated with each tone: high-level is a relatively high pitch; rising tone is a shift from a lower to a higher pitch on a single vowel; falling tone is a shift downwards; and dipping tone is a slight fall in tone followed by a rise. The forms in (9) illustrate the way in which tone can distinguish between otherwise identical words:

9 HIGH bā 'eight'
RISING bá 'to pull'
FALLING bà 'a harrow'
DIPPING bǎ grammatical marker for object

(The diacritics for the tones are from the Pinyin Romanisation system.) In addition, Mandarin has a 'neutral' tone, which occurs on unstressed syllables and has different level tone values (mid, half-low, low) depending on the tone of the preceding syllable. Tone languages differ with respect to the level tones and contour tones they use, and the extent to which lexically assigned tones are adjusted in the phonology.

A standard approach to tone systems is to treat tone as a specification of pitch height in terms of pitch segments; contour tones can be represented as the assignment of a string of two or more pitch height segments to a single vowel. The Mandarin words in (9) can thus be represented as:

```
10 bā   bá   bà    bǎ
   |    /\   /\    /|\
   H   L H  H L  M L H
```

(H = high pitch; L = low pitch; M = mid pitch)

Li and Thompson (1977) and Clumeck (1980) are early studies reporting the acquisition of tone. A recent review by Tsay (2017) compares the results of development of various Chinese languages. Tsay observes the following: infants acquire first level tones and then falling tones, followed by rising tones and more complex tones, and the acquisition of tones is completed well before the acquisition of the consonantal system. It is important to note that the children studied are as young as twelve months (the one-word stage, see Chapter 4) and that the methods used (for example, naming pictures) ensure that the researchers could be confident that the correct tones were identified.

As mentioned already, there are however more complex tonal systems – that is, systems that do not rely solely on the mapping between word and tone(s) but impose rules that effect changes in tone across words and morphemes. Demuth (1995) reports a study of Sesotho, one such language. She finds that learning the correct tonal rules and their application can take up to age three years (the oldest age reported for the child she studied). For example, Sesotho obeys the Obligatory Contour Principle (OCP), a constraint that blocks sequences of the same tone, by processes of delinking the tone segment from the word or morpheme. The child studied failed to obey the dictates of the OCP even at three years.

*The structure of syllables.* Syllables have an internal structure: the core of the syllable is the *nucleus*, which may be flanked by an *onset* and by a *coda*, as shown in (11), which parses the word *dunk* [dʌnk]:

11 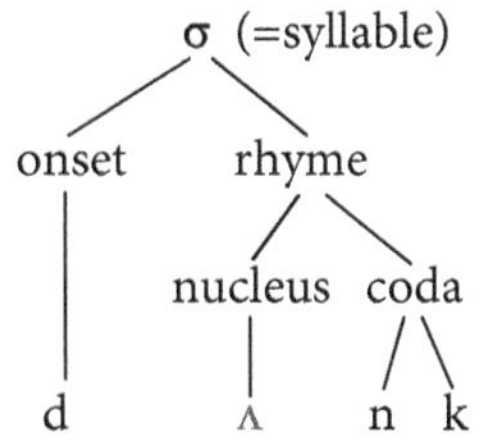


Cross-linguistically, it has been shown that CV (consonant-vowel) is the most frequently found syllable type, with some languages permitting only that. Other languages have a more extensive range of syllables, with some or all of the possibilities V, VC, CVC, CCVC, CVCC and CCVCC permitted.

Levelt et al. (1999) studied the development of syllables in Dutch, a language that permits the entire range of syllable types. The subjects were twelve children aged 1;0 to 1;11 at the beginning of the study (the data was drawn from Fikkert 1994 and Levelt 1994), i.e. the children were beyond the babbling stage and were producing recognisable words. They found an ordering in which the first four syllable types to be mastered were the same for all the children, followed by a divergence between two groups before the last, most complex syllable type was acquired, as shown in (12):

Group A: CVCC → VCC → CCV → CCVC<br>
12 CV → CVC → V → VC CCVCC<br>
Group B: CCV → CCVC → CVCC → VCC

Examples of the first stage (only CV syllables) from the data of Jarmo, aged 1;5, are given in (13):

| 13 Target | | Child | Gloss |
|---|---|---|---|
| /pus/ | → | [ pu ] | cat |
| /klar/ | → | [ ka ] | ready |
| /oto/ | → | [ toto ] | car |
| /api/ | → | [ tapi ] | monkey |

Thus, at the initial stage all children used only the universal syllable type. As shown in the two last examples in (13), the child had to add to the input to preserve the CV pattern.

*Sensitivity to stress and the acquisition of stress systems.* Stress – the perceptual prominence of a syllable or syllables in a word – is a complex matter acoustically, with several factors (length, intensity, pitch) contributing to our perception. There is no doubt that stress plays a role in the child's mispronunciations. A common child mispronunciation at the earliest stages is for unstressed syllables to be omitted (data from American English in Kehoe 2000, cited in Tessier 2016):[7]

14 Target Child
ba'by [ be ]
'kitty [ ki:]
sham'poo [pu: ]
alli'gator [ ge ], [ge:]

15 Target Child
'bubbles ['bʌbo]
gir'affe [va], [wa]
ba'nana ['nana]
e'lephant ['aIsə], ['æfə ]
alli'gator ['æde], ['æg3]

It is clear that the child prioritises stressed syllables, even when s/he is no longer confined (as in 15) to utterances of only one syllable. Moreover, when s/he produces bisyllables, the child chooses a pattern that conforms to [Sw] (=Strong-weak, Stressed-unstressed), i.e. s/he reduces a word such as *banana* to ['nana], rather than [ba'na]. The production of [Sw] syllables is found in several (but not all) child productions and is known as a *trochaic bias* (Allen and Hawkins 1980).

Allen and Hawkins proposed that the trochaic bias held for all child languages. Contrary evidence (in favour of a neutral start) has been argued for by Rose and Chamdoizeau (2008). Adult languages vary according to whether their stress system is *trochaic* (Sw) or *iambic* (wS), and child productions have been found to be sensitive to the nature of the input language. English and Dutch have a trochaic system, and K'iche' has an iambic system. Demuth (1996) draws on data from English, from Dutch (Wijnen et al. 1994) and from K'iche' (Pye 1992) to show that the simplifications children make are sensitive to the adult system of the language to which they are exposed. English preserves the stressed syllable (as shown in (14–15)), as does Dutch (16):

16 Target Child
'ziekenhuis 'hospital' ['sikhʌys]
'olifant 'elephant' ['o:xant]

The K'iche' child also preserves the stressed syllable, which, contrary to English and Dutch, is the final syllable:

17 Target Child
jo'lom 'head' lom
le'met 'bottle' met
chi'kop kop
wa'ik ik

However, there may still be evidence for a trochaic bias. Hebrew has about three quarters of words in child and adult speech with word final stress, which should favour an iambic system, leaving a quarter of words with non-final stress (Adam and

Bat-El 2009, as reported in Tessier 2016). Adam and Bat-El found that in the child's earliest productions, words with penultimate stress are truncated to the pattern Sw, and words with final stress are produced more variably, including the Sw pattern.

English has a trochaic system but does permit end stress. English stress is dependent on (1) the 'weight' of the final syllable, and (2) the syntactic category of the word. The weight of the final syllable depends on the number of consonants that follow the vowel and the quality of the vowel (roughly, whether the vowel is long or short). The exact specification of vowel length is a relatively complex matter. It is sufficient here to note that length distinguishes between pairs of vowels in English, such as the long vowel [i] (the vowel in *beet*) and the short vowel [ɪ] (the vowel in *bit*).[8] In English, a syllable is heavy if it has a long vowel and/or a final consonant.

Although the exact rules are complex, the broad effect of the weight of the word final syllable in English is to prevent stress from moving towards the middle of the word. For adjectives and verbs, the weight of the final syllable is calculated ignoring the last consonant. Stress is placed on the last syllable of a verb such as *usurp*, because the syllable has two consonants at the end, and is heavy even if the last consonant is ignored. Similarly, stress goes on the last syllable of the adjective *discreet* because the vowel in the last syllable is long, and a long vowel always means a heavy syllable. In a word such as *develop*, the vowel in the last syllable is short and is followed by only one consonant, and stress can skip over the last syllable, to the middle syllable.

Syntactic category adds complication to the system. For nouns, there is a tendency for stress to be placed one syllable to the left of where it would be placed for verbs and adjectives. Although there is a good deal of complexity in the system, the generalisation that stress goes to the left in nouns as opposed to verbs is a real one, and shows up clearly in contrasts such as that in (18):

18 con'vict 'convict
(verb) (noun)

How do children fare in learning the stress system of their language? We have seen that children are attuned to the stress patterns of the language, prioritising stressed syllables. Fast-forwarding to age seven, we can see that children have grasped the elements of the English stress system – in particular, the effects of syllable weight and word class described above. Smith et al. (1982) asked child and adult subjects to read aloud nonsense words such as those in (19), where the context sentence made the nonsense word unambiguously a noun or verb:

19 The *nuvit* was made in the factory (noun context)
The man had to *nuvit* the tractor (verb context)

All of the test words had two syllables; they had either a single consonant in final position (as in the 'nuvit' example above), or two consonants (for example, 'rafust'). Length of vowels is not represented in a consistent way in English, so whether the final vowel was interpreted as long or short was a matter of how the individual chose to say the word. Table 2.4 shows the percentage first-syllable stress according to

Table 2.4 Percentage first syllable stress

| | | | | Children | | Adults | |
|---|---|---|---|---|---|---|---|
| Number of final consonants | Quality of final vowel | Example: nonsense word | Example: similar English word | Noun | Verb | Noun | Verb |
| 1 | Short | nʌvit (nuvit) | edit | 89 | 78 | 85 | 50 |
| 1 | Long | nʌvi:t (nuvit) | discreet | 65 | 38 | 71 | 18 |
| 2 | Short | ræfʌst (rafust) | distrust | 87 | 56 | 50 | 25 |

*Note:* Adapted and abbreviated from Smith et al. (1982: table 4). The transcriptions are as given by Smith et al. The vowel transcribed as [i] in the nonsense word *nuvit* (parallel to the English word *edit*) would appear to be the vowel [I].

number of final consonants (one or two), quality of the vowel (short or long) and category of the word (noun or verb). The table shows that children have essentially the same pattern as adults. Both are more inclined to put stress on the first syllable if it is a noun rather than if it is a verb; if it has a short final vowel rather than a long final vowel; and if there is one consonant at the end of the word rather than two (for children, this last point holds only in the case of verbs). Thus, the seven-year-olds in this study had extracted from the words they heard the essential elements of the stress system of their language, although the contrasts are not as stark as they were for adults. Hochberg (1988) uses both experimental data and spontaneous speech to argue a similar point with respect to Spanish, using subjects as young as three.

## 2.7 CHILD SYSTEMS AND PHONOLOGICAL THEORY

In the past twenty years, many phonologists have advocated a shift in orientation from rule-based systems to constraint-based systems. In a rule-based system, an underlying form is changed to produce a surface form – i.e. what is pronounced. The derivations in (8) use a rule-based system, transforming the adult pronunciation into the child's utterance. In adult phonologies, rule-based systems use the phonological rules to demonstrate the difference between unpredictable, underlying forms and the actual pronunciation. As we saw above, the difference between the English words *cup* [kʌp] and *cut* [kʌt] must be represented in the lexicon of the language – the difference between the final sounds [p] and [t] is a meaningful difference and cannot be predicted. But the fact that the initial segment [k] is aspirated (pronounced with a breath of air) is predictable from the initial position of the sound and can be specified by a rule linking the base form to the actual pronunciation; the fact of aspiration does not have to be entered in the lexicon. A constraint-based system uses a ranking of constraints to effect the same result, by ranking a general constraint banning aspiration of consonants below a more specific constraint banning non-aspirated consonants in initial position. It is hypothesised that the constraints are universal; only the ranking of the constraints varies from language to language. The constraint-based theory of phonology is called *Optimality Theory* (OT), and is based on (*inter alia*) the work of Prince and Smolensky (2004).

A basic distinction in OT is the difference between *faithfulness* constraints and *markedness* constraints. Faithfulness constraints in a language limit the difference between an underlying form and its pronunciation. Markedness constraints look to the languages of the world to evaluate what is typical and what is atypical. As described in section 2.4, the notion of markedness is rooted in the ideas of Roman Jakobson (1968), although Jakobson did not use the terms 'marked' or 'markedness'. Jakobson observed repeated and frequent parallels between the development of language in the child and the frequency of phenomena across languages. A simple and bold idea in OT with respect to child language development is that markedness constraints have priority over faithfulness constraints, and development involves a reranking of faithfulness constraints, so that they come to dominate markedness constraints (Gnanadesikan 2004, and others in the same volume).

Gnanadesikan's daughter was aged 2;3 to 2;9 in the period that Gnanadesikan studied. At that time, the child exhibited a fairly steady state in her productions, and her productions were those typical of a child that age. She was learning standard American English. Gnanadesikan's analysis (like that of Smith 1973) takes the adult pronunciation as a basis for the child's output.

In Gnanadesikan's analysis a markedness constraint – *Complex – which bars any syllable of more than one initial consonant – outranked the (family of) faithfulness constraints, which requires that all input segments be present in the output. The child reduced all syllable initial consonant clusters to a single consonant, as shown in (20):

| 20 | Adult word | Child production |
|---|---|---|
| | clean | [kin] |
| | please | [piz] |
| | blue | [bu] |
| | draw | [dɔ] |

This result can be obtained by ranking the constraint *Complex over the constraint Faith. The ordering of constraints left to right indicates their ranking; the representation in (21) is known as a tableau and the * indicates a violation of the constraint. The arrow indicates the winning output, which violates the lowest-ranked constraint,

| 21 | | *Complex | Faith |
|---|---|---|---|
| | Clean: | | |
| | klin -> klin | * | |
| | klin -> kin → | | * |

Not all consonant clusters were simplified in the same way. *Clean* was simplified to [kin], with the second consonant eliminated, but *sky* was simplified to [gay], with the first consonant lost (all stop consonants were voiced, hence the change of [k] to [g] in *sky*). The choice of which consonant to eliminate was determined by the *sonority* of the segment. Sonority refers to the degree of obstruction in the vocal tract in the course of producing the consonant; the sonority hierarchy in (22) organises sounds from least sonorous to most sonorous (where > equals less sonorous):

22 stop > fricative > nasal stop > liquid > glide > vowel

In general, syllables are organised with the outer edges containing the less sonorous sounds, and with progressively more sonority as we get to the nucleus of the syllable, as illustrated in (23):

23 drunk: d r ʌ n k<br>
Stop – Liquid – Vowel – Nasal – Stop

The sequence [sk] violates this typical organisation of syllables, with a fricative preceding a stop. The child's change of [sk] to [g] reflects this: she chose as the preserved consonant the one that is less sonorous.

Vital to Gnanadesikan's analysis (and OT analyses in general) is the identification of adult languages that share the same constraints that the child has. The restriction to CV syllables is found in Cayuvava, Hua and Hawaiian (Tessier 2016). And Gnanadesikan observes (following Whitney 1889) that the pattern of preserving the least sonorant consonant is found in Sanskrit, where a reduplication process (a process of repeating a syllable) selects the least sonorant member of a consonant cluster for the copy (for example, when the perfect tense is formed, [prach] reduplicates as [pa – prach] and [s$t^h$a:] reduplicates as [ta – s$t^h$a:]).

## 2.8 ONGOING DEBATES

*Consonant harmony.* Although many child pronunciations are similar to the processes that operate in adult phonology, there is an exception to this parallelism: consonant harmony. The examples in (24a–j) are from the speech of Trevor, cited in Pater and Werle (2003), with additional examples (24k–l) from Trevor, cited in Becker and Tessier (2011):

| | | | | |
|---|---|---|---|---|
| 24 | a | [gɔg] | 'dog' | 1;5 (years; months) |
| | b | [kok] | 'coat' | 1;5 |
| | c | [kæ:g] | 'cat' | 1;3 |
| | d | [gɪ:gu:] | 'tickle' | 1;7 |
| | e | [gʌg] | 'bug' | 1;5 |
| | f | [kʌk] | 'cup' | 1;5 |
| | g | [gɪgu] | 'pickle' | 1;6 |
| | h | [bɛ:p] | 'bed' | 1;7 |
| | i | [bʌbə] | 'butter' | 1;7 |
| | j | [pap] | 'top' | 1;6 |
| | k | [tæt] | 'cat' | 2;0 |
| | l | [tan] | 'Tom' | 1;5 |

Examples (a–g) show a consonant becoming velar under the influence of another velar consonant in the word; examples (h–j) show a consonant becoming labial under the influence of another labial in the word; and examples (k–l) show a consonant becoming alveolar under the influence of another alveolar. Sometimes the agreement (the harmony) is regressive, where the second consonant influences the first (a, d, e, g, j and k) and sometimes it is progressive, where the first consonant causes the second one to change (b, c, f, h, i and l). Both front and back vowels can intervene, and harmony can occur across syllable boundaries (the harmonised pronunciations of *giggle*, *pickle* and *butter*). This type of harmony is very frequent in child language (see, for example, Smith 1973, 2010; Vihman 1978; Goad 1997), although the patterns vary from child to child, and yet it appears to be non-existent in adult languages. Adult languages do not exhibit the harmony of major class features (labial, alveolar, velar) that is characteristic of child consonant harmony (Shaw 1991).

An attractive idea about how to deal with child consonant harmony is in terms of under-specification for unmarked consonants. Rice and Avery (1995) propose prototypical representations for the stops /t/, /p/ and /k/. Modifying their representations to use the features we have been working with, we have:

```
25 /t/      /p/         /k/
    |        |           |
          -coronal    -coronal
                      -anterior
```

The idea is that the unmarked consonant /t/ (recall that /t/ is present in all languages, see p. 15) has no specification, whereas the more marked consonants /p/ and /k/ have specification for coronal and (in the case of /k/) anterior. This allows an explanation of the fact that alveolars are more frequently the target for consonant harmony than labials and velars (for example, Becker and Tessier 2011 report that the child, Trevor, whose harmony is reported in (24), had about three times the harmony when velars are harmonised to other segments than when alveolars are harmonised). The under-specification of /t, d/ allows the -coronal and -anterior specifications to spread to the /t, d/ segment:

```
26 d   ʌ   k
   |       |
        -coronal
   ↑    -anterior
   └─────
```

Thus, the under-specification account allows the seemingly odd fact that the unmarked consonant is more often targeted for consonant harmony than more marked consonants. However, harmony that targets /k, g/ and /p, b/ and spreads /t, d/ does exist, and, moreover, child consonant harmony persists at the point at which the child can produce the full range of harmonised sounds. It remains an open question as to whether the child has a non-adult rule of phonology, or whether the child operates under the influence of schemata that govern production (see for example, Macken 1992).

*Types of markedness and trajectories of learning.* In a study of the speech of the child, Trevor (24, above), Becker and Tessier (2011) observed the following asymmetry: the learning curve for consonant harmony was a U-shape, as shown in (27a); but the learning curve for syllable initial consonant clusters was an S-shape (always simplifying a cluster such as /tr/ in *train* to /t/, followed by a period of variable performance, and then ultimate mastery), as shown schematically in (27b),

27 a

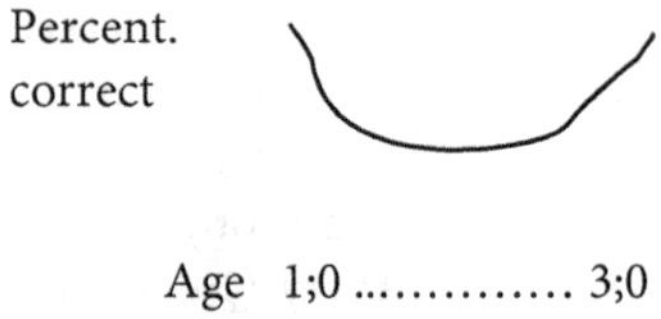


27 b

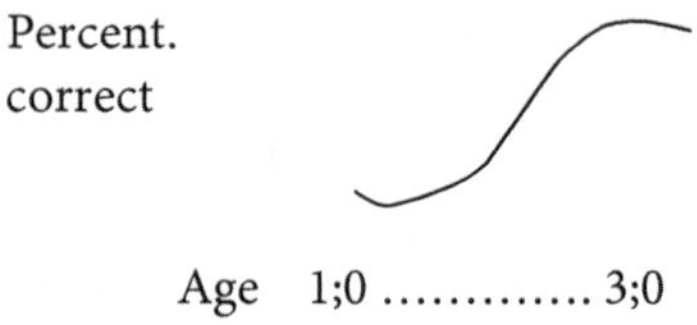


The difference in learning curves echoes a difference in languages of the world: consonant harmony is never attested in child's form in adult languages, whereas avoidance of initial clusters is very frequently found in adult languages.

*The nature of the child's lexicon.* Menn (1983) and Menn and Matthei (1992) argued for a dual-lexicon approach to children's representations: a permanent store and a store that was accessed in the child's productions. Becker and Tessier (2011) offer an account that resembles the two-lexicon approach as a base for modelling the learning path towards adult competence. A review of results from, *inter alia*, aphasia and neuroanatomy (Gow 2012) suggests that this may be a fruitful way to pursue that development of the child's lexicon.

*Rule ordering.* Smith (1973) argued that the child's phonological rules must apply in a particular order – for example, in (7) above, rule 3 must apply before rule 6, since rule 3 provides a context for the velarisation of the stop, and this context is destroyed by rule 6. The thrust of recent phonological theory has been to eliminate rule ordering, and the role of perceptual limitations has eaten away at the motivation for rule ordering in child languages. But there is no inherent need for the new emphasis on constraints to lead to the abandonment of rule ordering. In principle, constraints and rule ordering can co-exist. Earlier work on adult phonologies argued forcefully for the need for rule ordering, and data such as that presented in Exercise 2 below offers a challenge for a theory of child phonology that eliminates rule ordering.

## CHAPTER SUMMARY

This chapter has sketched the development of sound systems. The ability of infants in the first year of life undergoes rapid development, the discrimination of vowel sounds having a head start over discrimination of consonants. Nonetheless, by the one-year mark both vowels and consonants are attuned to the distinction made in the ambient language. The development of phonology reflects structures in adult languages, albeit these may not be present in the language the child is exposed to. The young child is sensitive to properties of the input, particularly in the areas of tone and stress. The recent emphasis in Optimality Theory on constraints has led to predictions concerning orders of acquisition, with unmarked structures having a developmental priority. Consonant harmony remains a problem, in that consonant harmony is frequently found in child language but is never attested in adult systems in the form found in children.

## FURTHER READING

Jakobson (1968) is a classic reference and well worth reading. This chapter has focused on the development of child phonologies in terms of what does and doesn't occur. Dresher (1999) gives a more detailed account and contrasts the problem of what children actually do, with the problem of how the child gets from the input data to the adult system (the developmental problem vs. the logical problem of acquisition). Tessier (2016) is a textbook devoted to Optimality Theory in child phonology.

## QUESTIONS AND EXERCISES

1. Is there a way of linking the early skill in recognising vowel sounds found by Kuhl et al. (1992) and the early mastery of tones reported by Tsay (2017)?
2. The pronunciation of certain forms by two brothers aged four and five (Applegate 1961) deviated systematically from the pronunciation of their parents, who spoke the American English dialect of their community. We can state two rules to account for the boys' pronunciation:

   Rule 1: Change [s] to [t] or [d] (depending on a voiced or voiceless sound elsewhere in the word).
   Rule 2: Change [t,d], [p,b] and [k,g] to a glottal stop if any of these sounds is identical to another sound in the word.

   (ʔ is a glottal stop, heard for example in Cockney British English pronunciation of the [t] in words such as *bottle*; [ɨ] is centralised [i]; [y] is the palatal glide, [j]). Those readers with a course in phonology can rewrite these rules more elegantly in terms of phonetic features. How must these rules be ordered to produce the correct results?

| Adult word | Child pronunciation |
|---|---|
| walks | wɑkt |
| talks | tɑkt |
| talked | tɑkɨʔ |
| toot | tuwʔ |
| suit | tuwt |
| kick | kIʔ |
| pet | pɛt |
| bit | bIt |
| tag | tæg |
| died | dayʔ |
| does | dad |
| takes | teykt |

(The data in this problem is from Halle and Clements 1983.)

3. How would you go about handling the data in Exercise 2 in a constraint-based system? I.e. what would you look for in support of a constraint-based analysis?
4. What do you think about drawing general conclusions about children's phonological abilities on the basis of cases such as that in Exercise 2, where the children behaved in a way that was special – not typical of other children's development?
5. The child studied by Smith (1973) showed the following stages in the production of consonant clusters with an initial [s] (1973: 63–64; some stages have been omitted):

   Stage A: all preconsonantal /s/'s disappear.
   Stage B: /sr/ clusters appear, but [s] does not come from /s/.
   Stage C: /st/ clusters appear.

   Discuss these stages in the light of the sonority hierarchy sketched in the text (see p. 26) and in the light of the child's ability to produce sequences of sounds present or not present in the ambient language.
6. Hochberg (1988) used an imitation task to test children's knowledge of the stress rules of Spanish. She asked children to imitate non-existent words that either conformed to the rules of Spanish stress or violated them. The basic finding was that children made more errors in imitating nonsense words with non-regular stress, than nonsense words with regular stress. Find some English-speaking children and try out her task with materials such as those used in the Smith et al. (1982) experiment described in the text. That is, ask children to repeat sentences with nonsense words in noun and verb contexts, with stress on the last syllable or the first syllable. For example:

   Noun context: The [ 'nʌvit / nʌ'vit ] was made in the factory.
   Verb context: The man has to [ 'nʌvit / nʌ'vit ] the tractor.

   Vary the weight of the final syllable of the nonsense word both in terms of the number of consonants and the length of the vowel.

## NOTES

1. Do not worry about understanding the physics of speech; it is sufficient in what follows that you understand the concepts that the experimental results are built around.
2. The mel scale is based on pitch and can be converted to hertz.
3. The cues for speech perception are complex, and the perception of voicing is no exception. Perception of voicing may be manipulated in relation to the acoustic structure of adjacent vowels. In stop-plus-vowel sequences, voicing perception is dependent on the onset of the first (lowest) formant of the vowel relative to the second and third formants – a relatively early onset signalling voicing and a later onset, after the second and third formants, signalling voicelessness.
4. Gruber expresses the generalisation in terms of the feature [+/−grave], a feature motivated by the fact that labials and velars share some acoustic properties that distinguish them from alveolars (Jakobson et al. 1952).
5. See Pater, Stager and Werker (2004) for further discussion.
6. It is the convention in phonology to use slashes to enclose pre-final forms and square brackets to enclose phonetic output.
7. The stress mark (ˈ) precedes the stressed syllable.
8. It is of issue how quality correlates with and contributes to characterising a vowel as long or short. The phonetic feature *tense/lax* (greater or lesser constriction of the tongue root and tongue body) frequency correlates with length, tense vowels being longer. A vowel is also long if it is diphthongal.

## CHAPTER 3

# MORPHOLOGICAL DEVELOPMENT

One of the adages of language acquisition studies is that acquisition is a creative process. Yet the thrust of studies summarised in the last chapter would seem to be that the opposite is true. Some of the results reviewed there suggest that constraints found in adult phonologies also place bounds on the possible analyses a child can make. We will see in Chapters 4 and 5 a similarly close correspondence between child and adult systems in many areas of syntactic development. Overall, language development can be seen as a highly constrained process, with quite strict limits on the types of analysis a child can entertain.

Where, then, is the creativity? A plausible answer is that the examples most frequently cited as evidence of creativity are innovative word forms. Morphological rules in the adult grammar account for the formation of new words (innovations in the vocabulary) and for the outer shape of existing words in particular contexts (*one dog* but *two dogs*, with the plural *-s* occurring after *two*). Novel forms created by children strike the ear as such and give evidence that the child is active in using her/his grammar to produce words not in the adult vocabulary; we will see several different types of example below. Nonetheless, the child's innovations appear to adhere to a system of morphology that is governed by constraints also found in the adult grammar. Thus, we see that the child's 'creativity' with respect to word formation in fact offers quite firm evidence that child grammars are constrained by principles that also govern adult systems.

### 3.1 INFLECTIONAL MORPHOLOGY

Morphological rules are traditionally divided into two types: *derivational* and *inflectional*. The distinction between derivational and inflectional rules has been drawn in various ways. But a basic difference is that rules of inflectional morphology do not create new words. Instead they regulate the form of the word according to the syntactic context in which it occurs. Thus in English we have rules that require a progressive verb form to be marked with *-ing* (indicating ongoing activity), a plural noun to have an affix *-s*, an *-s* also marks possession, the past tense of a verb has the affix *-ed*, and there must be an *-s* on a present tense verb that has a third person singular subject (*he/she/it* in example (1)):

1 Progressive: I am walk*ing*
Plural: One dog (singular) – two dog*s* (plural)
Possessive: John'*s* book
Past tense: I walk (present) – I walk*ed* (past)
Third person agreement: I walk – he/she/it walk*s*

The inflectional morphology of English is limited in complexity compared to other languages: the rules just listed are about it. Moreover, there are exceptions to the rules. Nouns such as *child, tooth* and *mouse* have the plurals *children, teeth* and *mice,* not *childs, tooths* and *mices;* and verbs such as *know, see* and *think* have as a past tense form *knew, saw* and *thought,* rather than *knowed, seed* and *thinked.*

### 3.1.1 THE ACQUISITION OF ENGLISH INFLECTIONAL MORPHOLOGY

One of the ground-breaking studies in the last fifty years was that of Brown (1973). Brown looked at the acquisition of fourteen morphemes by three children. The morphemes were a heterogeneous set, including prepositions (such as *on* and *in*), as well as inflectional elements. Brown found that there was a fairly regular order of development, and that those regular inflections included in the study were not particularly early acquisitions. Table 3.1 gives the ages at which seven of the morphemes were mastered (where mastery was defined as correct use in 90 per cent or more of the contexts in which the morpheme was obligatory). As the table shows, although the most precocious of the children (Eve) had mastered all but one of the morphemes by age two, one (Sarah) was almost five years old before she mastered the regular past tense ending.

As the English child learns to command the regular rules of the language, s/he makes mistakes of overuse of them. Children's errors strike the ear – the landlord of a local bar corrected his young son when the boy used *seed,* reminding him that he should have said *saw.* Attentive to his father's wishes, the boy changed his utterance: 'I sawed it!' Study of the pattern of use of regular and irregular verb forms has revealed that over-regularisation errors are in fact very rare – never more than 5 per cent of the total usage of a verb – and occur only after the regular rule had been acquired (the data for Adam, who acquired the regular past tense rule at age 3;6 is given in Marcus et al. 1992, and summarised in Pinker 1995).

Table 3.1 Age of mastery of seven morphemes

| | Age of mastery (years;months) | | |
|---|---|---|---|
| Morpheme | Adam | Sarah | Eve |
| Present progressive *-ing* (John is walk*ing*) | 2;6 | 2;10 | 1;9 |
| Preposition *on* | 2;6 | 2;10 | 1;9 |
| Preposition *in* | 2;6 | 2;10 | 1;9 |
| Plural marker (two book*s*) | 2;6 | 2;3 | 1;11 |
| Possessive marker (John'*s*) | 3;2 | 2;10 | 1;11 |
| Past regular *-ed* (John walk*ed*) | 3;6 | 4;10 | 1;11 |
| Third person singular (John walk*s*) | 3;6 | 4;0 | 2;3 |

### 3.1.2 ENGLISH VS. INUKTITUT: A DIFFERENCE SO FAR LARGELY UNEXPLAINED

If a system with few inflectional markings can take up to five years to master, how would a child do with a much more complex system? Surprisingly, children learning languages such as Greenlandic show remarkable prowess with the system. Fortescue (1984/5) reports a study of a two-year-old child's acquisition of that language. The study was a small one, based on just half an hour of recorded speech during which the child was playing with his mother. Greenlandic has a great number of productive affixes: 'A typical word consists of a stem followed by from zero to at least eight derivational affixes then an obligatory inflectional ending … all bound together by complex morphophonemic patterns of morpheme attachment and fall under one potential intonational tone unit' (Fortescue 1984/5: 101–102). During the half-hour recorded speech, the child produced forty separate inflectional endings as well as twenty-four derivational affixes (Fortescue reports that there are 318 inflectional and over 400 derivational endings in Greenlandic). The child's longest single-word utterances are given in Table 3.2.

Perhaps the child studied by Fortescue was – like Brown's Eve – a precocious learner. However, other studies have shown a consistent pattern of ease in learning affix systems that are 'transparent', i.e. with each affix having a particular meaning and occurring in a particular sequence, with no coalescence. Crago and Allen (2001) studied several children learning Inuktitut (spoken in Canada, and from the same language family as Greenlandic Eskimo). Some examples of child use of affixes in the study by Crago and Allen are given in Table 3.3; all of the children are under three years of age. Crago and Allen argue that the use of verbal affixes by the children is productive – i.e. not restricted to fixed phrases. These children's very early knowledge of the verbal inflection appears to be typical of the acquisition of languages with transparent morphological systems (see also Aksu-Koç and Slobin 1985 for Turkish, and Clancy 1985 for Japanese).

Table 3.2 Single-word utterances of Aqissaq, a two-year-old speaker of Greenlandic Eskimo

| | |
|---|---|
| Utterance: | tattuus-sinnaa-nngil-angut |
| Morpheme-by-morpheme gloss: | be crowded-can-not-1st person plural indicative |
| Meaning: | 'We cannot be (so) crowded together (in it)' |
| Utterance: | nangia-ssa-nngil-anga |
| Gloss: | be scared-future-not-1st person singular indicative |
| Meaning: | 'I shan't be scared' |
| Utterance: | uppi-ti-qa-akkit |
| Gloss: | fall-cause-begin-intensifier-1st/2nd singular indicative |
| Meaning: | 'I'm going to make you fall!' |
| Utterance: | anartarfilerisu-u-pput |
| Gloss: | sewage collector-be-3rd person plural indicative |
| Meaning: | 'They are the sewage collectors' |

*Source:* Adapted from Fortescue (1984/5: 198, table 2).

Table 3.3 One- and two-word utterances of children learning Inuktitut

| | |
|---|---|
| Utterance: | Nirilangannginama (Paul 2;11) |
| Breakdown of morphemes: | niri-langa-nngit-gama |
| Morpheme by morpheme gloss: | eat-FUT-NEG-CVS-1sS (FUT = future; NEG = negation; CVS = causative; 1sS = first person singular subject) |
| Meaning: | 'I won't eat' |
| Utterance: | Piipiapimik tigumialutit (Sarah 1;11) |
| Breakdown of morphemes: | piipi-apik-mik tigumiaq-lutit baby-DIM-MOD-SG hold-ICM-2sS (DIM = diminutive; MOD = modalis; SG = singular; ICM = incontemporative; 2sS = second person singular subject) |
| Meaning: | 'You're holding the baby' |
| Utterance: | Anaana aarqitait? (Elijah 2;0) |
| Breakdown of morphemes: | anaana aarqik-jait mother fix-PAR.2sS.3sO (PAR = participative; 2sS = second person singular subject; 3sO = third person singular object) |
| Meaning: | 'Mom, did you fix it?' |

*Source:* Crago and Allen (2001).

Several lines of thought have been offered for the contrast between English vs. languages such as Inuktitut and Turkish with respect to the mastery of inflectional morphology. However, these candidate explanations do little more than repeat the facts of the languages involved (see for example, Aksu-Koç and Slobin 1985: 874, who appeal to semantic transparency as one factor), and the different speeds of acquisition for languages with impoverished vs. rich inflectional systems seem to me to be largely unexplained.

### 3.1.3 ROOT INFINITIVES

Another example of the importance of cross-linguistic data is the phenomenon that has come to be known as a *root infinitive*. As we have seen, English has a poor inflectional system in the present tense, with only the third person singular being marked (*s/he sings* vs. *I/we/you/they sing*); thus only in the third person can a tensed verb be distinguished from an infinitive (*to sing*). In Table 3.1, we saw that this inflection is the last to be acquired. Other languages have more developed systems, marking person (first, second, third) and number (singular, plural) more consistently. This permits us to see more readily patterns in the use of inflection. A basic finding is that in some languages, but not perhaps others to the same extent, children use an infinitive form of the verb as the main verb (a root infinitive), something that is ordinarily ungrammatical in the adult language.

There have been various accounts of this phenomenon (including Wexler 1998 and Boser et al. 1992), of which one of the most successful is that of Hoekstra and Hyams (1998). Hoekstra and Hyams assembled the results from studies of spontaneous production of inflection on verbs by children aged approximately eighteen months to thirty-six months in a range of European languages. Hoekstra and Hyams found that the alternation between use of inflection and use of a root infinitive was not arbitrary. In the examples in (2), from a child speaking Dutch, the child uses the infinitive form such as *kopen* rather than a tensed form such as present tense *koop*, as required in the adult language. Hoekstra and Hyams argue that root infinitives were used for predicates that denoted an event (such as *play*, *make* and *buy* in the examples in (2)), not a state:

2 a Niekje buiten spelen
Niekje outside play-infinitive
'Niek (=speaker) wants to play outside'
b Jij helicopter maken
You helicopter make-infinitive
'You must build a helicopter'
c Eerst kaartje kopen
first ticket buy-infinitive
'We must first buy a ticket!'

When the verb denoted a state (e.g. verbs meaning *want*, *please*, *need* and so on), then the verb was inflected for tense. In addition to the requirement that root infinitives denote an event, root infinitives have a modal interpretation – as indicated by the interpretations given in (2), although the effect is more nuanced in some languages (Hyams 2011). Hoekstra and Hyams suggest that the use of root infinitives when the verb denotes an event and the modal interpretation of root infinitive utterances are connected. The particular modal meaning associated with root infinitives is deontic, i.e. associated with the necessity or desirability of a future event, and evidence from adult languages points to deontic modality being found in combination with predicates that denote events. It fits with the lack of a dedicated infinitive form in English that there is an absence of restrictions on when bare forms are used in child English – the use of bare forms is not limited to eventive predicates and intended modal meanings. Some examples of English child speech using non-eventive verbs are given in (3) (from Hyams 2011):

3 a Becky have puzzle
b The baby want a bottle
c Ann need Mommy napkin

The analysis of root infinitives given by Hoekstra and Hyams is far from uncontroversial, but it serves to illustrate the facts to be accounted for.

## 3.2 DERIVATIONAL MORPHOLOGY

Derivational rules are processes which form new words, including processes for forming words by adding affixes or by joining words together to form a compound word. For example, in English we can derive a noun from a verb by adding the affix *-er*, which carries an agentive meaning:

4 shave (verb)
shave + er → shaver (person or thing who shaves)

Different parts of speech can also be joined in a compound:

5 kill (verb) + joy (noun) → killjoy
towel (noun) + rack (noun) → towel rack
sky (noun) + blue (adjective) → sky blue
red (adjective) + hot (adjective) → red hot

There is an internal structure to English compounds. We can assign the hierarchical structure for the example *killjoy* in (6):

6
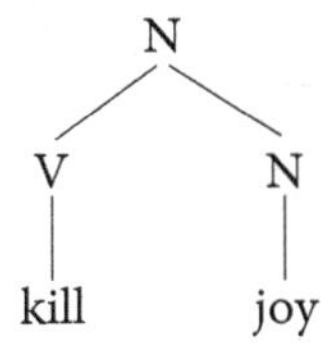


The category of the compound word is determined by the rightmost element; thus *killjoy* is a noun (N) and *sky blue* is an adjective (A).

### 3.2.1 INNOVATION IN THE ACQUISITION OF DERIVATIONAL MORPHOLOGY

There are two types of innovation in derivational morphology: innovation in the use of affixes to form new words, and innovation in using words in a particular manner, for example, the use of a noun as a verb.

An example of the first type is given in Clark (2001). The child Clark studied made a distinction between the adjectives ending with *-y* and *-ed*; examples are given in Table 3.4. English does not make a distinction between adjectives used to describe permanent properties and those used to describe temporary states, but some languages, such as Spanish, do. The verbs *ser* and *estar* both translate as 'be', with *ser* being used before adjectives that express a permanent, inherent state and *estar* being used before adjectives that express temporary states.

The second type of innovation has also been studied by Clark (1982). She studied the child's production of *denominal verbs* – verbs formed by using a noun as a verb, such as *to garden* and *to table* (a decision). She found that children innovate such forms with a variety of meanings, most frequently instrument and locatum (in locatum verbs, the noun that is used as a verb denotes an entity that is being placed somewhere). Some examples from English-speaking children are given in Table 3.5 (similar innovations are made by French- and German-speaking children). While denominal verbs are quite richly documented in children's spontaneous utterances,

Table 3.4 Inherent vs. temporary states in innovative adjectives

| *Inherent state adjectives in -y* | *Child D, age* |
|---|---|
| It isn't crumby [= full of crumbs, speaking of Amaretti biscuits] | 2;6 |
| It's very nightly [= dark, when being driven home] | 2;7 |
| There's a rocky house [= house made of rocks] | 2;10 |
| *Temporary state adjectives in -ed* | |
| My foot is all crumbed [= the bottom of the foot covered with crumbs] | 2;6 |
| That fork is all ... all BUTTERED [= covered in butter] | 2;6 |
| These are all floured [= covered in flour after mother has just covered pieces of veal with flour] | 2;6 |

*Source:* Adapted from Clark (2001: table 13.3).

Table 3.5 Examples of instrument and locatum denominal verbs

| *Instrument* | *Age of child* |
|---|---|
| You always have to scale it first (wanting to have some cheese weighed) | 2;4 |
| I broomed her (having hit his sister with broom) | 2;7 |
| Is it all needled (asking if the pants his mother is mending are ready) | 2;3 |
| *Locatum* | |
| Mummy trousers me (talking about getting dressed) | 2;3 |
| I'm crackering my soup (putting crackers in her soup) | 3;11 |
| Will you chocolate my soup? | 5;0 |

Table 3.6 Innovative causatives

| | *Age of child* |
|---|---|
| It always sweats me (= makes me sweat) | 4;3 |
| This is aching my legs (= making my legs ache) | 5;3 |
| Enough to wish me one of those beds (= to make me wish for ...) | 5;8 |
| Do you want to see us disappear our heads (= make our heads disappear) | 6;0 |

*deverbal nouns* (for example, *the shave* meaning 'lather') are observed much less frequently, matching the observation of grammarians that the process of forming nouns from verbs has always been less frequent than that of forming verbs from nouns (Marchand 1969, cited by Clark 1982: 418).

Sometimes children's morphological innovations exploit devices that are quite marginal in the particular language that the child is learning. Bowerman (1982) found that at a certain stage her daughters innovated causative forms of verbs – using verbs with the intended meaning that the subject of the verb causes the object to undergo the action denoted by the verb. Some examples are given in Table 3.6. These examples are striking because they show children experimenting with a morphological device widely used in languages of the world, but confined in English to a relatively small number of alternations (for example, *the door opened* (non-causative) vs. *he opened the door* (causative))

### 3.2.2 CHILDREN'S KNOWLEDGE OF COMPOUNDING

In the adult grammar, compounds of the form N + [$_N$ Ver] are interpreted as having the first N undergoing the action of the second N: *rat catcher* means someone who catches rats. Note that in the case of a noun that is regular, i.e. forms the plural with the affix *-s*, only the singular form is permitted as the first N in the compound: **rats catcher* is ungrammatical. When an irregular noun is used to form the compound, either the singular or the plural can be used; both *mouse catcher* and *mice catcher* are acceptable.

Gordon (1985b) tested three- to five-year-olds' ability to produce N + [$_N$ Ver] compounds formed with regular and irregular nouns as the first N. Other studies indicate that the *-er* agentive rule and compound formation are within the abilities of three-year-olds (Clark and Hecht 1982). In Gordon's experiment, the child was shown sets of objects corresponding to both regular and irregular nouns – for example, a string of beads (regular: *bead*, *beads*) and some teeth (irregular: *tooth*, *teeth*). Knowledge or lack of knowledge of these irregular forms was established for each child. The child's task was to tell the experimenter if s/he thought a puppet, Cookie Monster, would like to eat the set of objects before her/him. Then the experimenter asked the child: 'What do you call someone who eats X?' (using in the position X the plural form that the child had previously supplied).

The children's responses to questions of the form, 'What do you call someone who eats X?', gave strong support of knowledge of the distinction in use of singular and plurals as the first N. In almost every case (161 out of 164 instances), children used the singular form in producing compounds with regular plurals. Thus, they produced forms such as *bead eater*, but not forms such as **beads eater*. And those subjects who demonstrated that they knew the correct form for irregular nouns also freely used that form in forming compounds: 36 out of 40 compounds involving irregular words by children who knew the irregular plural were of the form *teeth eater* as opposed to *tooth eater*.

As Gordon argues, it really is implausible that the children's behaviour can be put down to rules that they work out on the basis of the speech around them. Although either the singular or the plural of irregular nouns is allowed as the first element of a compound in the adult grammar (*mouse eater* or *mice eater*), the singular sounds somewhat more natural. Gordon inspected compound forms in a corpus of written English (Kučera and Francis 1967) and found that singular forms were used as the first element of compounds in virtually all instances of compounds formed with irregular nouns in the corpus (151 out of 152 instances).

But could it be that children in Gordon's experiment had a rule of affix stripping: there is an affix to strip (*-s*) in the case of regular plurals, but not in the case of irregulars in English. A study by Clahsen et al. (1992) argues against the affix-stripping hypothesis. German has five ways of forming plurals: adding *-e* (*der Hund* 'the dog', *die Hunde* 'the dogs'); -0 (zero) (*der Daumen*, 'the thumb', *die Daumen* 'the thumbs'); *-er* (*der Wald* 'the forest', *die Wälder* 'the forests');[1] *-(e)n* (*die Frau* 'the woman', *die Frauen* 'the women'); *-s* (*das Auto* 'the car', *die Autos* 'the cars'). The *-s* plural ending is argued to be the regular (default) ending for adult German speakers.

Table 3.7 Inclusion and omission of plurals in German-speaking children's compounds

| Compounds with -*n* | | Compounds with irregular affixes | |
|---|---|---|---|
| Included | Omitted | Included | Omitted |
| 17 | 17 | 10 | 0 |

*Source:* Adapted from Clahsen et al. (1992: table 10, summarising the performance of five children who both overgeneralised (*e*)*n* and produced compound nouns.)

In their spontaneous speech, some of the children in Clahsen et al.'s study over-regularised the -(*e*)*n* ending for plurals; examples are given in (7):

| 7 | indianer-n | 'indians' | (correct adult form *Indianer*) |
|---|---|---|---|
| | vogel-n | 'birds' | (correct adult form *Vögel*) |
| | Pferd-en | 'horses' | (correct adult form *Pferd-e*) |
| | zähn-en | 'teeth' | (correct adult form *Zähn-e*) |
| | bätt-en | 'leaves' | (correct adult form *Blätt-er*) |
| | aut-en | 'cars' | (correct adult form *Auto-s*) |

Critically, when these children used compounds, the over-regularised ending -*n* was omitted some of the time; the non-regular endings were never omitted. Table 3.7 summarises their production of compounds, and (8) gives examples of the incorrect omission of the -*n* affix:

| 8 | tanne-bäum | 'christmas tree' | (correct adult form *Tanne-n-baum*) |
|---|---|---|---|
| | blume-vase | 'vase' | (correct adult form *Blume-n-vase*) |
| | 'auer-hof | 'farm' | (correct adult form *Bauer-n-hof*) |
| | küche-fenster | 'kitchen window' | (correct adult form *Küche-n-fenster*) |

These German children's productions are evidence that the children did not employ a simple strategy of affix stripping, rather they stripped only the suffix they had overgeneralised.[2]

### 3.2.3 WHERE IS THE THEORY?

The experimental evidence just summarised argues that young children are sensitive to the distinction between regular and irregular nouns when they form compounds. What is the basis for this sensitivity? A common point in the proposed analyses is that irregular nouns are listed as such in the lexicon (the mental entries for words) of the language, whereas regular nouns require some extra rule or mechanism to apply (see among others, Kiparsky 1983; Siddiqi 2006). An analysis by Harley (2011) links the distinction children draw to the process of *incorporation*. Compounds of the *bead eater* type are formed by incorporating the N *bead* to the left of the head noun *eater*. Incorporation is the combination of elements which produces a structure that is impervious to the processes that affect the larger structure in which the compound is embedded. Once incorporation has taken place, features such as the [plural] feature are no longer required to be 'checked'.[3] Feature-checking is a device that legitimises the structure: all features much be checked for the structure to be allowed and incorporation blocks that checking mechanism; hence regular plurals are blocked within the compound. Why is the plural of irregulars allowed? This is because the plural of irregulars is an independent entity in the lexicon which can compete with the singular form for entry into the structure, with no feature-checking necessary.

### 3.2.4 MORE ON CHILDREN'S COMPOUNDING

We end this chapter with an example that is on the boundary between morphology and syntax. Snyder (2001, 2007: chapter 5) explores a connection between verb particle constructions in which the verb and the particle can be separated (such as *set down*, in sentences such as *He set the box down*), and root compounds, i.e. compounds that consist of two or more bare (unaffixed) words. Examples of root compounds are *alarm clock*, *frog man*, *force feed* and *age old*. Although root compounds are found in most languages, the ability to innovate meanings is restricted to only a few languages, including English. The compound *frog man* has a lexical meaning in English (underwater diver), but English speakers can freely extend the meaning of the compound – for example, to a man who does scientific research on frogs, to a man who collects frogs or to a man who looks like a frog. In Spanish, by contrast, only the lexicalised meaning is allowed: *hombre rana* 'man frog' can only mean an underwater diver.

In a cross-linguistic survey of languages of different families, Snyder found that novel bare root compounds were a prerequisite for the existence of separable particle verbs. Languages with both bare root compounds and separable particle verbs existed, and languages with only bare root compounds existed, but no language had separable particle verbs without having bare root compounds. Snyder tested if this asymmetry held for children aged around two years learning English, with the prediction that children should not acquire particle verbs before bare root compounds. This prediction was borne out: the age of the First Repeated Use[4] for bare root compounds and particle verbs were found to coincide statistically.

## CHAPTER SUMMARY

This chapter has summarised the remarkably constrained, yet creative, morphological abilities of young children. Children learning languages such as Inuktitut, with a complex but transparent morphological system, very rapidly get to grips with the inflectional system of their language. For other language types, the system may take some time to master, but nonetheless the child's errors are systematic: root infinitives are limited to certain verb types, and innovative compounds are sensitive to the child's default plural. We can take this combination of innovation and constraint to be evidence that the child is working with a system in which the broad outlines are set and must be followed.

## FURTHER READING

Brown (1973) is a classic study.

## QUESTIONS AND EXERCISES

1. Berko (1958) invented the 'wug' test. Presented with one creature, the child was told that it was a wug. Next, s/he was shown a picture of two of the same creatures and asked what they were. The reply 'wugs' gave evidence that the child had a grasp on the regular plural in English. Can you think of a way in which a wug test could be used to investigate plural formation in German?
2. Clark et al. (1986) report that children learning compounds such as *man hugger* go through three stages in the process of development. First, they only produce compounds in which they use no affixes, resulting in forms such as *hug man*. At the second stage, they begin to add the *-er* affix, but still do not use the adult order, with the result being compounds such as *hugger man*. Finally, children begin to get the order correct and produce the adult form *man hugger*. What may influence the child in the first two stages? How could a cross-linguistic survey provide support for your answer?

## NOTES

1. The *-e*, -o and *-er* endings may also include a change of vowel quality in the stem.
2. It should be noted that the children in Clahsen et al.'s (1992) study were diagnosed as *dysphasic*, a term used by Clahsen and colleagues for children who have problems with subject and verb agreement processes, and with case assignment on nouns (where nouns occur in different form depending on their function in the sentence). These deficits appear to be orthogonal to their performance in Clahsen et al.'s study.
3. Feature-checking/Agree is a mechanism used in recent syntactic theory (see Chapter 4).
4. First Repeated Use is a measure invented by Snyder that allows the researcher to eliminate isolated uses from consideration (see Snyder (2001, 2007) for details).

# CHAPTER 4

# THE ACQUISITION OF SYNTAX

This chapter summarises (some of) the basic facts about the nature of syntactic systems, together with the results of language acquisition studies. The descriptive framework is roughly that of the 'Government and Binding' theory in Chomsky's book *Lectures on Government and Binding* (1981) and subsequent works; this framework – although almost forty years old – gives us an organisational handle on the nature of syntactic systems, and is one that has been used in many of the language acquisition studies we will summarise. We turn to developments in the theory of syntax since the early 1990s (the Minimalist Program) in later sections (4.4.1 and 4.6).

## 4.1 SYNTACTIC STRUCTURES AND UNIVERSAL GRAMMAR

### 4.1.1 BASIC SYNTAX

Several concepts are basic to the description of syntactic systems. Words are assigned to syntactic categories: noun, verb, adjective, preposition and others to be mentioned shortly. These words head or 'project' phrases – thus a noun heads a noun phrase (NP), a verb heads a verb phrase (VP), and so on. These syntactic phrases organise the linear strings of words that make up a sentence into a hierarchical structure. Thus, the English sentences (1) and (2) will have the approximate structures shown in (3) and (4):

1 Sue left

2 The judge gave the verdict

3

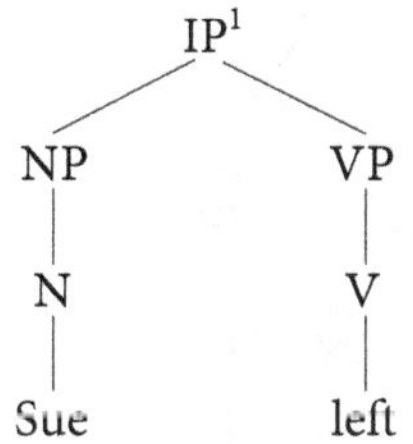


IP = Inflectional phrase
NP = Noun phrase
VP = Verb phrase
N = Noun
V = Verb

4

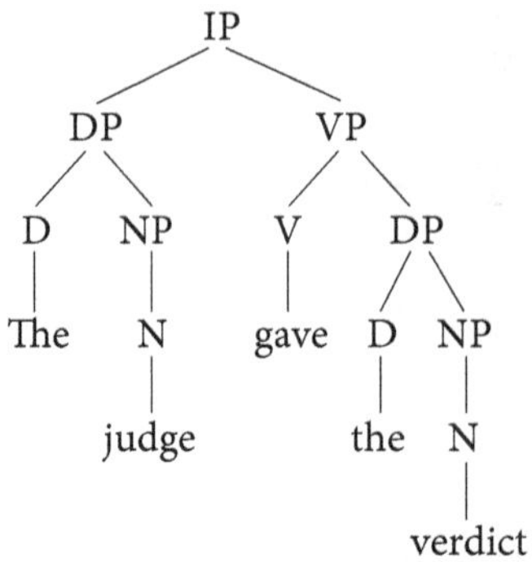


DP = Determiner phrase
D = Determiner

Sentences and phrases may be embedded within one another, producing structures of potentially infinite length. For example, sentences can be stacked inside a verb phrase, as illustrated in (5a–c) and (6):

5 a Sue said that Horace sighed
  b Sue said that Hilda thought that Horace sighed
  c Sue said that Hilda thought that Alice said … that Horace sighed

6

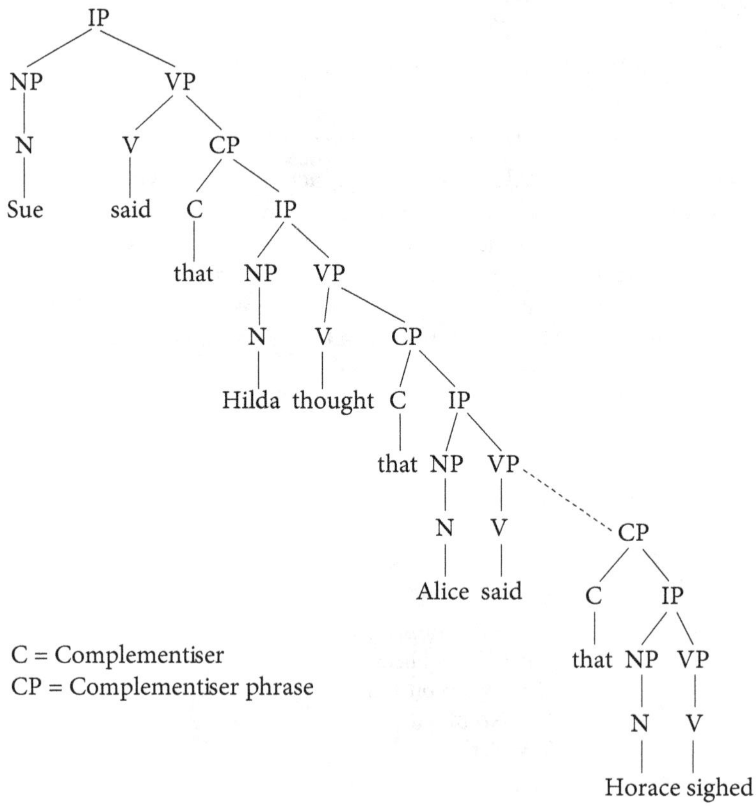


C = Complementiser
CP = Complementiser phrase

Similarly, relative clauses follow the noun they modify in English, and again can be stacked up to build potentially infinite structures:

7 a Dogs that gnaw on bones
  b Dogs that gnaw on bones that fall from trucks
  c Dogs that gnaw on bones that fall from trucks that drive into town that …

Languages differ in their basic syntax. Languages such as English, with a subject-verb-(object) order in simple sentences are *right-branching* – the head of the phrase is generally on the left and modifying structure is built up to the right, as illustrated by the structure in (6). Languages such as Japanese, with a subject-(object)-verb basic order in simple sentences, are *left-branching*, building up modifying structure to the left of the head. Thus, a complement sentence is placed to the left of the verb it modifies in Japanese, and the pattern of embedding in the VP will be broadly the mirror image of that for English, as shown schematically in (8):

8

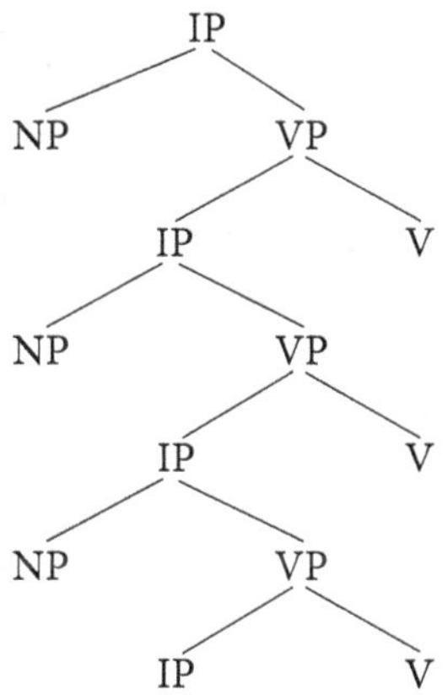


And in Japanese, relative clauses also precede the noun they modify.

Languages such as English and Japanese are frequently called *configurational* languages, because they organise the linear string of words into hierarchical structures. Other, so-called *non-configurational* languages, have 'flatter' structures and permit almost free variation of word order within sentence units. The Australian language Warlpiri is an example of a non-configurational language (see, for example, Hale 1983 for a description of some of the properties of Warlpiri). The division between configurational and non-configurational languages is not always clear-cut, and even non-configurational languages are held to build hierarchical structure at some level of abstraction (see Legate 2003). Nor is the organisation of phrases perfectly regular in all languages – for example, in English, adjectives precede the noun they modify, contrary to the general right-branching pattern of the language. But the general division between configurational languages, with relatively rigid word orders and a characteristic type of hierarchical structure (left- or right-branching), and non-configurational languages, with free word order, is valid as a first approximation and represents a basic distinction for which the theory of syntax must account.

### 4.1.2 EMPTY CATEGORIES AND MOVEMENT

Almost all modern work in syntactic theory recognises that sentence structures involve 'invisible' (or 'inaudible') parts: syntactic positions that are not fleshed out with words. Positing such invisible, empty categories aids, *inter alia*, in the explanation of the way we understand some types of ambiguities and relations between non-adjacent elements. For example, the title of an article by the psycholinguist Richard Cromer:

9 Children are nice to understand

is ambiguous: it can mean either 'It is nice to understand children' or 'It is nice of children to be understanding'. This ambiguity can be represented if the rules of English syntax permit two different structures to be imposed on the string in (9). On the first reading given above there will be an empty object position of the embedded verb, to which the main clause subject *children* is linked, as shown in (10a), accounting for the fact that *children* is understood as the object of *understand* ([e] in the structure represents an empty category). On the second reading, the main clause subject is linked to the empty subject position of the embedded sentence, as shown in (10b), accounting for the fact that *children* is understood as the subject of the verb *understand*:

10 a Children are nice [e to understand [e]]

b Children are nice [e to understand]

Several types of sentence, not merely ambiguous sentences, give rise to the need for empty categories. Questions are a prime example. In English, the questioned element appears at the front of the sentence and may correspond to a variety of types of phrase and positions in the sentence structure that follows. Positing empty categories gives us a way to account for these relations. We can represent the fact that the question word *what* in (11a) is understood as the object of the verb *drink* by positing an empty object position, to which the question word is linked:

11 a What will Frances drink?

b What will Frances drink [e]?

Similarly, for an object placed in initial position for emphasis:

12 a Drano, she drank!

b Drano, she drank [e]!

In short, we can account for facts about our understanding of sentences with an order that deviates from the basic order of the language by reference to a more abstract structure in which that basic order is represented. In Chomskyan generative grammar, the linkage in (11) and (12) is represented in terms of a movement operation. In the case of (11), two elements are shifted from the position they would occupy in a declarative sentence to form the question: *what* is moved from object position of the verb and *will* is moved to the left of the subject:

13

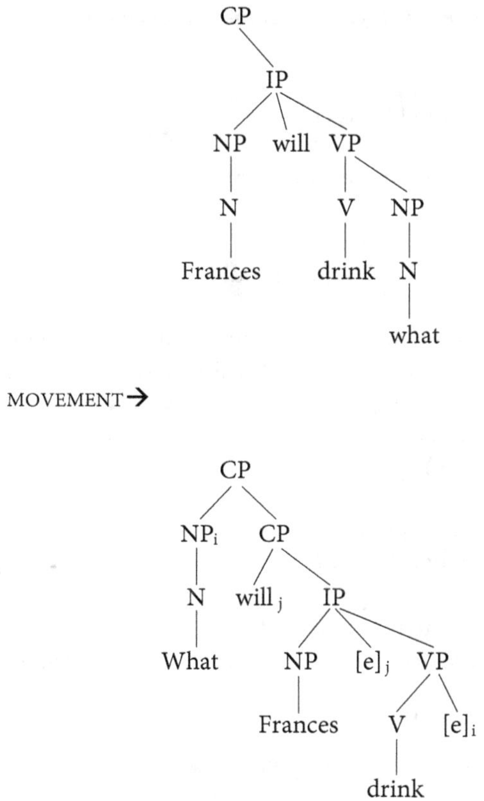


As can be seen in (13), movement is to the Complementiser Phase (CP) in this case. The movement operation is subject to structural restrictions, as we will see below (section 4.4.2).

Not all empty positions are created by movement. For example, the empty subject position of some types of embedded sentences is represented by an abstract pronominal element: PRO. We have already seen a sentence with such an empty position: a more exact structure for the reading of (10) (*Children are nice to understand*) in which *children* is taken as the logical object of *understand* is one in which there is a PRO subject of the embedded verb; thus (10a) can be recast as follows:

14 Children$_i$ are nice [PRO to understand [e]$_i$]

with the subject PRO having no specified reference.[2] In other examples, PRO must refer to a specific NP inside the sentence. This will be the case for the alternative reading of (10), in which *children* is the subject of the verb *understand*, and for examples such as (15):

15 My aunt promised [PRO to leave]

where the subject PRO of *leave* is interpreted as *my aunt*. The interpretation of the subject PRO in (15) does not change, even if we embed the sentence inside another sentence:

16 Lucinda said my aunt promised [PRO to leave]

Although there is no logically necessary reason for that – grammar aside, (16) might mean something like (17):

17 Lucinda said my aunt promised that she (Lucinda) could leave

But it does not. The reference of PRO is determined by rules that place an index on the NP to which it must refer. Thus, *my aunt* and PRO will receive the same index in (15). Similarly, co-indexation is used to represent co-reference relations between overt pronouns and their intended referent, such as the definite pronoun *him* in (18) and the reflexive pronoun *himself* in (19):

18 Lucifer claimed that the angel deceived him
19 Lucifer claimed that the angel deceived himself

In (18), *him* may refer to *Lucifer* but not to the angel; in (19) *himself* must refer to *the angel* and not to *Lucifer*. These rules are dependent on syntactic structure, as we shall see in section 4.4.1.

### 4.1.3 LEXICAL DEMANDS

Individual words can place restrictions on the structure of a sentence. The lexicon of a generative grammar contains much of the sort of information that we find in an ordinary dictionary: an indication of the pronunciation of a word and of its meaning. In addition, there is information about the local syntactic environments into which the word can be inserted – its *subcategorisation(s)*. Thus, for example, a transitive verb such as *devour* subcategorises for a direct object NP; an intransitive verb such as *dive* does not:

20 a John devoured a banana
   b *John devoured
21 a *John dived a bellyflop
   b John dived

Subcategorical restrictions limit the phrase categories that can be immediately adjacent to a word – thus a verb can in general impose subcategorical restrictions that occur with it directly under the VP, but not on the internal structure of the subcategorised phrases. Nor do subcategorical restrictions generally extend to the subject NP. In addition, recent theory proposes that verbs may be marked with features that determine the syntactic derivation that they undergo, as we will see below (section 4.4.1).

The lexical entry of a word will specify the semantic or *thematic* roles associated with the phrases with which it combines, including the subject NP. Thus in (21b), the subject of *dived* has the thematic role 'agent', but if we change *dived* to *died*:

22 John died

the subject NP is no longer an agent but has the thematic role of *experiencer* or *theme*. There is controversy about the exact set of thematic roles needed in grammatical descriptions, but the set is frequently taken to include *agent*, *patient*, *experiencer*, *theme* (thing acted upon or affected, perhaps subsuming patient and experiencer), *goal* (recipient or end point with respect to physical or mental transfer) and *location*.

The role of notions such as 'subject' and 'object' in grammatical theory is highly controversial. However, it is widely thought that the grammatical relation subject (or 'external argument' in the terminology of Williams 1981) has some special status, and the thematic role assigned to the subject position is frequently marked as such in the lexical entry of verbs.

### 4.1.4 UNIVERSAL GRAMMAR

A prominent feature of Chomskyan syntactic theory since the early 1980s has been an emphasis on the formulation of general principles, from which the properties of particular grammatical phenomena will follow. The most frequently given illustration of this is the fact that whereas in earlier-style grammar (for example, Chomsky 1957, 1965) there were a great many different rules for individual constructions – for example, a rule of question formation, a rule for topicalising phrases and a rule for producing passive syntax – in more recent theory this is reduced to a single general operation: movement. Whether a language uses movement, what can move, and where to, is motivated and restricted by the dictates of principles of grammar.

The form and nature of these principles and the range of variation in their execution is a matter dealt with by the theory of *Universal Grammar* (UG). Some principles will be absolute. For example, phrases may move to a higher position in the syntactic structure, but not to a position lower in the structure ('height' may be defined in terms of the c-command relation detailed in section 4.4.1 below).[3] Other principles may take on a limited set of values, accounting for observed differences in human languages. For example, the theory of phrase structure must permit both head-first and head-final languages and must allow also for free word order languages. These are the broad parameters of differences in phrase structure. This general approach to UG is termed the 'principles and parameters' approach.

It should be clear from the above that UG is not equivalent to 'things true of all languages', although this error of interpretation persists in the literature. Rather, UG is construed as a set of principles that limit the variation between languages.

### 4.1.5 MODULES IN THE THEORY OF GRAMMAR

On the view developed in Chomsky (1981) and later works, syntactic theory comprises several interacting modules, each of which constrains a particular type of grammatical entity. The modules of grammar that we will be most concerned with below are:

*X-bar (X') theory.* X' theory is the theory governing phrase structure configurations. X is to be construed as a variable ranging over the various syntactic categories (N, V, P, A[djective]). The term *bar* or *prime* (X') refers to layers of structure posited within a phrase, for example:

N'' = NP<br>
|<br>
N' = intermediate level<br>
|<br>
N = word level

Since the intermediate phrasal level is not crucial in what follows, with one or two exceptions only the word and phrase levels are given. (And such an intermediate level in recent theory is produced only when the mechanisms of the grammar require the level.)

*Binding theory.* Binding theory deals with restrictions on the co-reference of anaphoric elements, including definite and reflexive pronouns.

*Bounding theory. Bounding* theory deals with restrictions on the operation of movement.

*Control theory.* Control theory is the theory determining the referential possibilities for the empty (PRO) subject of clauses.

Each of these subtheories is a semi-autonomous module, with its own principles. However, these principles draw in many instances on the same concepts. For example, 'height' (defined in terms of c-command, see section 4.4.1) enters into principles of binding theory, bounding theory and control theory.

### 4.1.6 THE ROLE OF SYNTACTIC THEORY IN LANGUAGE ACQUISITION

A child's acquisition of the syntax of her/his language should be, to put it grossly, a matter exactly as complex and as simple as the (correct) theory of syntax. That is, s/he must find out which type of language is being learnt (what the parameter settings for the language in the various modules of the theory are) and, in doing so, the child should be aided by the fact that such variation is limited and that general principles and concepts traverse the different modules, reducing the complexity of the system.

A common-sense first guess at how syntax acquisition progresses might be to say that the child first must set the general configurational pattern of the language, and then may take some time to sort out the details of the system within the separate modules. Although many of the studies of acquisition summarised in the following sections were not carried out in the context of the theory of Government and Binding sketched above, we can use their results to piece together a picture that is largely compatible with this view of syntactic development as 'basics first, details later'.

## 4.2 THE OUTER COURSE OF DEVELOPMENT

Before beginning a topic-oriented account of syntactic development, it is worth summarising some basics about the outer course of language development – what kind of utterances a child tends to produce at what ages. In the early 1970s, many detailed studies of children's speech were carried out (see, for example, Bowerman 1973; Brown 1973; papers in Ferguson and Slobin 1973). These studies systematically recorded the language of one or more children, supplementing older diary studies (such as Grégoire 1937, 1947). The 1980s saw the beginning of the computerisation of corpora of child speech with *The CHILDES Project*, directed by Brian MacWhinney. The project – which is still ongoing – makes available speech samples of many languages in transcribed form and sometimes audio- and video-recorded (see MacWhinney 2000). The upshot from both older and newer resources is that there are some striking regularities in language development, with the same or very similar patterns of development being found for different children in different homes and environments. However, there is also considerable variation between children, making the task of establishing 'stages' a challenge.

The age that a child achieves a particular level of development varies from child to child (Brown 1973, see table 3.1). But there are general age guidelines that we can indicate. At around the turn of the first year, children begin to produce one-word utterances – that is, single words that are for the most part recognisable words in the adult vocabulary. By twenty months, the child has a vocabulary of about fifty words (Nelson 1973) and enters a 'two-word' stage, combining words together, although not always in sequences that are well formed in the adult language. Children's early multi-word utterances are frequently referred to as 'telegraphic speech', since children learning languages such as English tend to omit the sorts of words (determiners such as *the* and *a*, auxiliary verbs, prepositions) that we leave out when writing a telegram (for those born in the last two or three decades and unfamiliar with sending telegrams, attempt to rewrite the sentence *In questions, try to avoid the subjunctive* in three or four words). By the end of the third year, the child may be producing a range of complex sentence types (complements to verbs, adverbial clauses and relative clauses) and a four-year-old frequently gives the impression of being a fully fluent speaker of a language comparable to the language s/he is learning, if not identical to it. The reader can get some sense of the kinds of utterances and sentence types found in early child speech by studying Table 4.1, which gives data from the early multi-word speech of two children, and Table 4.2, which shows a typical order of emergence of various sentence types.

Table 4.1 Early speech of two children

***Speech of Gregory*** (Braine 1963)
From age 19 months to 22 months:
31 combinations with *byebye* (e.g. *byebye plane, byebye man, byebye hot*)
14 combinations with *see* (e.g. *see boy, see sock, see hot*)
5 combinations with *allgone* (*allgone shoe, allgone vitamins, allgone egg, allgone lettuce, allgone watch*)
5 combinations with *it* (*do it, push it, close it, buzz it, move it*)
3 combinations with *my* (*my mummy, my daddy, my milk*) and *big* (*big boss, big boat, big bus*)
2 combinations each with *pretty* (*pretty boat, pretty fan*), *nightnight* (*nightnight office, nightnight boat*), *hi* (*hi plane, hi mommy*) and *more* (*more taxi, more melon*)
20 unclassified combinations (e.g. *mommy sleep, milk cup, oh my see*)

***Speech of Abigail*** (MacWhinney 2000, Wells corpora files 1, 2 and 3)
Age 18 months (approx.):
8 two-word utterances (e.g. *a bang, this way, baba mummy?*)
6 three- or more word utterances (*do it for me, I want mummy, I said there, what's that*)
Age 21 months (approx.):
23 two-word utterances (e.g. *no out, no mummy, cut it*)
8 three- or more word utterances (*this cut it, I cut it, goes on there, what is that?, this is a boot* =[boat])
Age 24 months:
26 two-word utterances (e.g. *bike mummy, where book?, and car*)
29 three- or more word utterances (e.g. *then we'll play, goodbye piano tuner, ball of wool, Mummy must have gone shopping, having my lunch*)

## 4.3 EARLY SYNTAX

Determining the exact nature of the child's earliest syntactic system is an extraordinarily difficult task. But there is evidence that fairly early on – certainly by the third year – children do have a system that conforms in basic ways to the syntactic patterns of the language that is being learned.

### 4.3.1 THE BEGINNINGS OF PHRASE STRUCTURE

If we take a look at the utterances listed for the two children in Table 4.1, the first child Gregory shows some degree of repetition in his patterns of speech. He has thirty-one combinations with the first element *byebye*, and fourteen with the first element *see*, and other combinations that occur less frequently. This led Braine (1963) to propose a grammar in which certain words had a special status, which Braine dubs pivots. Such pivot words can occur with many other words. The preference for first elements as pivots in two-word utterances (there is a similar preference in the two other children Braine studied) suggests that these utterances are the first attempts of the child to utter structures that conform to the head followed by complement/modifier pattern of English phrase structure. Thus, the so-called telegraphic

Table 4.2 The emergence of sentence types

| Approximate age | QUESTIONS | | CO-ORDINATION | EMBEDDING | |
|---|---|---|---|---|---|
| 18–20 months<br>One-word utterances | | | | | |
| 20–24 months<br>Telegraphic speech | Yes-no questions signalled by intonation only:<br>Fraser water?<br>See hole?<br>No eat?<br>Wh-questions with the form Wh + N or Wh + NP V;<br>Question words generally limited to *what* and *where*, and use only a narrow range of verbs:<br>What(s) that?<br>What cowboy doing?<br>Where Anna pencil?<br>Where horse go? | | | | |
| Third year | *Why* and *Why not* questions; *What* questions with a wider range of verbs:<br>Why you smiling?<br>Why not he eat?<br>What the dollie have? | | Co-ordination by juxtaposition: You lookit that book; I lookit this book | V + V sentences, often with *wanna* ('want to'), *hafta* ('have to') and *gotta* ('got to') as the first verb: I hafta peepee<br>Infinitival complements with the infinitive marker *to*:<br>I want to go<br>Embedded wh-complements:<br>I know when to go | |
| | Yes-no questions with some auxiliary verbs and some inversion of the subject and auxiliary:<br>Will you help me?<br>Can I have a piece of paper?<br>Can't it be a bigger truck? | Wh-questions with some auxiliary verbs, but frequently without inversion of of the subject and auxiliary:<br>Why kitty can't stand up?<br>Which way they should go? | Co-ordination with *and*: He was stuck and I got him out | Subordinate clauses introduced by *if*, *so*, *'cos* and *when*:<br>I want this doll because she's big<br>Some types of relative clauses:<br>… ones Mommy got?<br>… ball that I got? | Tensed complements to verbs:<br>I guess/think she's sick |

*Source:* Based primarily on Limber (1973) and Klima and Bellugi (1973).

speech may represent the child's first steps in a language that conforms to the patterns of the adult input. The speech patterns of the second child Abigail are less repetitious, in the sense of using particular words in 'pivot' position, and she shows a steady obedience to the structure of English phrases – her utterances can generally be converted to 'correct' English by inserting words. And by twenty-four months of age, she is capable of such sophisticated utterances as *Mummy must have gone shopping*. That utterance was surrounded by what appears to be practice stabs and follow-ups, shown in (23) (there were no intervening utterances in the transcript):

23 Mummy (xxx, unintelligible)
yes Mummy
shop
Mummy must have gone shopping
gone shopping
gone shopping
gone

### 4.3.2 SUBJECTLESS SENTENCES

One characteristic of children learning English is their tendency to omit subjects. In some languages, such as Italian, this is grammatical. But in English, we cannot say *gone shopping*, except in special circumstances. A subject is ordinarily obligatory. Do children learning English go through a phase in which they misconstrue the grammar, setting the parameter to the value appropriate for a language of the type exemplified by Italian? This was the speculation of Hyams (1986). A great deal of research has followed. An early cross-linguistic study of children speaking English vs. children speaking Italian showed that subject omission is at a higher level in Italian (Valian 1990). And a study by Bloom (1990) showed that English-speaking children were sensitive to the structure of sentences in which they omitted subjects: they omitted subjects when the VP was longer. Moreover, some languages such as Hebrew have a mixed system, with fixed and second person pronouns optional and third person pronouns obligatory in the past and future tenses; data from Hebrew shows that very early on (by two years) the basic patterns of the adult language have emerged (Levy and Vainikka 2000).

Such observations are compatible with the hypothesis that English-speaking children had the parameter set to the English value but omitted the subject due to pressures on an immature system for language production. Rizzi (2008) argues that children learning English and other languages that do not permit the subject to be dropped indeed *do* have the correct setting for their language, and do not permit subjects to be omitted freely. However, children in the null-subject stage (which extends until the middle of the third year) have a grammar that is slightly different from the grammar of adult English. Roeper and Rohrbacher (1994), for example, observed that children permit only the subject of main clauses to be omitted, and they do not permit omission of subjects in wh-questions, when the verb is marked for tense.

This is reminiscent of oral dialects of English such as that studied by Thrasher (1977, reported in Rizzi 2008), who found the following pattern:

24 a (I) thought I heard something
b I thought *(I) heard something[4]
c __ can't do it, can I/you/he/she/they/we?
d More problems, *(I) don't need
e What do *(you) want?

Rizzi concludes that children speaking English have the correct setting of the parameter for English. However, he does consider the role of performance mechanisms, a topic which we will take up further in Chapter 7.

## 4.4 SYNTAX IN THE PRESCHOOL YEARS

This section deals with the syntactic knowledge of children aged approximately two to six or seven years. Once the basics of the system are in place (whether the language is right- or left-branching), it makes sense to ask about the development of operations and principles whose application depends on the basic structures. We will look at knowledge and development in each of the modules listed in section 4.1.5: binding theory, bounding theory and control theory.

### 4.4.1 CHILDREN'S KNOWLEDGE OF THE BINDING THEORY

Binding theory is concerned with the referential properties of various types of pronouns. Some of the strongest results in child language in the late 1970s and the 1980s came from studies of children's interpretation of pronouns.

The binding theory can be given in a simplified form in terms of three principles, deriving from Chomsky (1981):

*The binding theory*
Principle A: An anaphor must be bound in its local domain
Principle B: A pronominal must be free in its local domain
Principle C: An R-expression must be free

The crucial terms to understand are: *anaphor*; *pronominal*; *R-expression*; *bound* and *free*; and *domain*.

An 'anaphor' is a pronoun of the reflexive type, such as *himself* in English, for which there must always be a co-referential NP in the same sentence. A 'pronominal' is a pronoun of the type of definite pronouns such as *he* or *him* in English, which may or may not refer to an NP in the same sentence (i.e. it may refer to an entity – mentioned or unmentioned – in the discourse environment). 'R-expression' is an abbreviation for referring-expression, and for our purposes it will mean a noun phrase such as *John*, *the boy*, *the government*, *a girl that I know*, etc. An element is 'bound' if it is co-indexed to another element that is at the same height or higher in the syntactic structure. If something is not bound, it is free. Height can be defined in terms of the relation *c-command* (short for constituent-command). The following definition of c-command derives from Reinhart (1976):

*C-command* Node A c-commands node B if and only if the first branching node above A dominates B and neither A nor B dominates the other.

In the schematic tree below, node Q c-commands R, S and T, node R c-commands node Q, node S c-commands T, and T c-commands S. In other words, a node c-commands all its sister nodes and all the nodes dominated by its sisters:

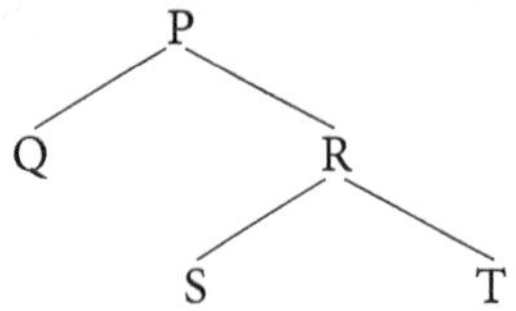


Finally, 'domain' refers to the structural space in which the principles operate; in the examples that follow (25–39), we can take the domain of an element (anaphor or pronominal) to be the IP node that is most immediately above that element.

We will look at some of the facts which are accounted for by the binding principles and then at children's sensitivity to the relevant distinctions in the adult grammar, taking first Principle C (which does not require reference to the notion of domain), and then Principles A and B.

*Principle C.* Principle C accounts for facts such as those illustrated by (25a–b):

25 a He said that John would leave before noon
b John said that he would leave before noon

In (25a) the pronoun *he* and the noun *John* may not be co-referential, but in (25b) such coreference is possible (although not obligatory). The basic generalisation is that a pronoun and a full noun phrase cannot co-refer when the pronoun is in a structurally dominant position with respect to the noun phrase. The structure for (25a) is given in (26):

26

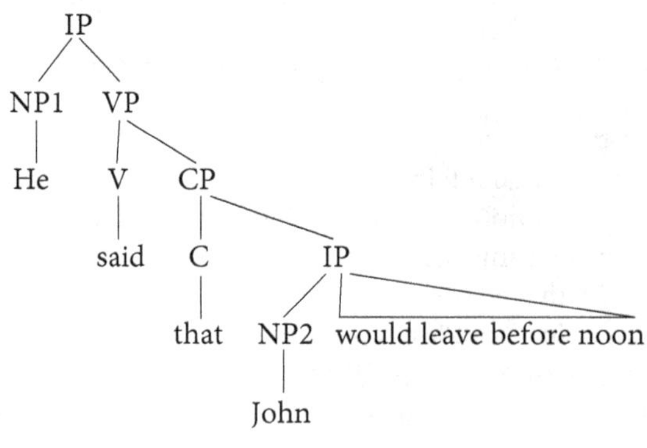

It is easy to see that the top NP (NP1) is in a structurally superior position to the subject of the lower clause (NP2). Structural superiority can be defined in terms of c-command: NP1 c-commands NP2. Principle C accounts for the ungrammaticality of (25a), on the interpretation where *he* and *John* are co-referential: NP2 (*John*) is an R-expression; if it is co-indexed with an NP that c-commands it (NP1) it is bound, but Principle C requires that an R-expression must be free (not bound) and so under Principle C co-indexation of NP2 to NP1 is barred. If the positions of the pronoun and the referring expression are reversed, as in (25b), co-indexation can take place with no violation of the principle: the lower NP (NP2) is then a pronoun, not an R-expression, and can be bound to the higher NP (NP1).

Several early studies showed children to be sensitive to the facts accounted for under Principle C (Solan 1978, 1983; Lust et al. 1980; Crain and McKee 1985). Lust et al. and Solan used an experimental task invented by Chomsky (1969), in which children act out their understanding of sentences using dolls and other props. One or more dolls not mentioned in the sentence are made available to the child, which the child can use if s/he wishes to act out an interpretation in which the pronoun refers to an entity not mentioned in the sentence. When the pronoun precedes the NP to which it could potentially refer, children have a strong tendency to take the option of making the pronoun refer to an entity outside the sentence. But, crucially with respect to knowledge of the structural restrictions on reference imposed by Principle C, children take the option of making the pronoun refer *inside* the sentence most frequently when sentence-internal reference is permitted under Principle C. Thus in Lust et al.'s study, which tested preschool and young school-age children, there was an average of 14 per cent of sentence-internal responses (reference between the pronoun *he* and the other NP in the sentence) for sentences such as (27), where Principle C blocks co-reference, compared with 23 per cent of such responses for sentences such as (28), where Principle C permits co-reference:

27 He turned around when Snuffles found a penny
28 When he closed the box, Cookie Monster lay down

If we look at the structures for these sentences, we can see that the lower portion of sentence-internal co-reference responses (14 per cent) is for the sentence type in which co-reference involves linking an R-expression to a c-commanding pronoun:

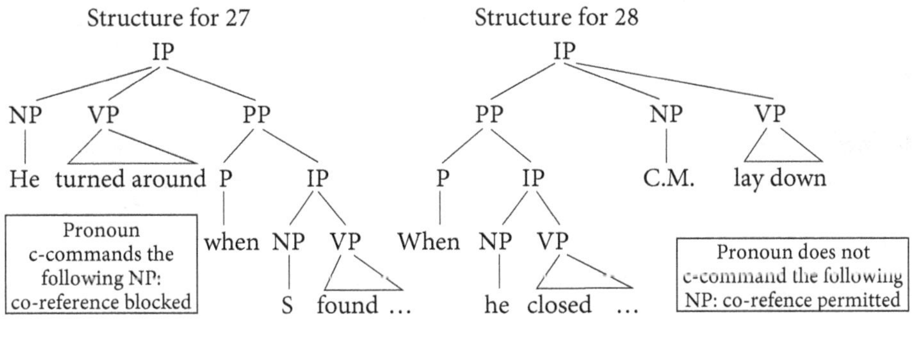

Solan's studies provide evidence that children are paying attention to c-command and not to some alternative structural restriction. For example, the results with sentences of the types (27) and (28) above could be accounted for if children blocked co-reference when the first IP node (as opposed to the first branching node) above the pronoun also dominated the noun phrase to which the pronoun is to be made co-referential. Exercise 1 at the end of this chapter shows the sentence types Solan used to argue that c-command as defined in this chapter is the basis for Principle C. Solan tested five- to seven-year-old children, and a study of three-year-olds by Goodluck and Solan (2001) confirms that c-command is – as far as we can tell – the child's first hypothesis concerning the structural restrictions on pronominal reference.[5]

These early results have been extended in experiments with other languages, including Russian (Kazanina and Phillips 2001) and Thai (Deen and Timyan 2018), which present more complex situations than English. For example, while adult speakers of Thai block co-reference between a pronoun subject and the subject of the subordinate clause in accordance with Principle C in sentences such as (29), adults speakers allow violations of Principle C when the R-expression is unmodified and the binder is an identical R-expression (as in 30), but obey Principle C when the R-expression forms part of a phrase containing an adjective (31). Thus, co-reference is blocked in (29), allowed in (30), but blocked again in (31):

29 *kháw$_i$ phûut wâa ʔaacaan$_i$ càʔ chanáʔ<br>
he say that teacher will win<br>
'He$_i$ says that the teacher$_i$ will win'

30 ʔaacaan$_i$ khit wâa ʔaacaan$_i$ càʔ chanáʔ<br>
teacher think that teacher will win<br>
'The teacher$_i$ thinks that he$_i$ will win'

31 *ʔaacaan$_i$ khon ʔûan$_i$ khit wâa aacaan khon ʔûan$_i$ càʔ chanáʔ<br>
teacher classifier fat thinks that teacher classifier fat will win<br>
'The fat teacher$_i$ thinks that he$_i$ (=the fat teacher) will win'

The Thai children were tested with a truth judgement task, in which the child had to verify whether a story that involved co-reference between two entities could be described with sentences that were the equivalent of (29–31). The children rejected all three sentences, unlike Thai adults who were willing to accept sentences of type (30). Thus, young Thai speakers had a strong form of Principle C, which later they had to modify to allow co-reference between the entities in (30). (See Deen and Timyan 2018 for discussion of what may lie behind the adult violation of Principle C in 30.)

*Principles A and B.* Principles A and B of the binding theory account for the difference in the distribution of NPs (DPs) to which definite pronouns and reflexive pronouns may refer. We saw this in sentences such as (18) and (19), repeated here as (32a–b):

32 a Lucifer claimed that the angel deceived him<br>
b Lucifer claimed that the angel deceived himself

The reflexive pronoun must refer to the subject of the lower clause (*the angel*), whereas the definite pronoun may not refer to that NP. If the definite pronoun refers to an entity inside the sentence, it must refer to the higher subject (*Lucifer*). Principle A of the binding theory requires that a reflexive must be bound in its domain and Principle B precludes binding a definite pronoun in its domain. Taking the IP node immediately above the pronominal element to define the domain of the element, we can see from the structure for (32a–b) that the principles will produce the right results for the co-reference facts:

33

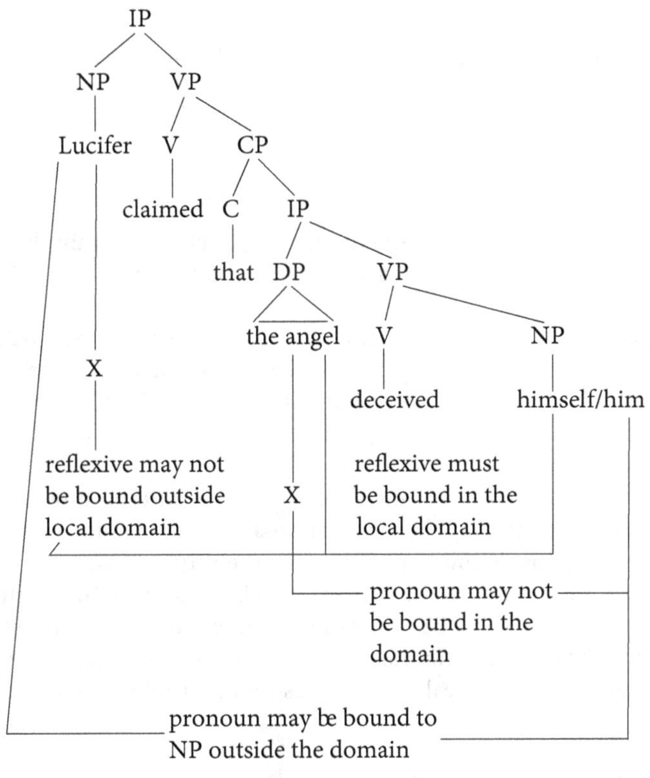


Multiple studies from the 1980s showed that young children are sensitive to basic distinctions dictated by Principles A and B. Moreover, their choice of a local antecedent for a reflexive pronoun is not simply a matter of choosing the linearly closest NP. When faced with the task of selecting a referent for the reflexive in sentence pairs such as (34a–b):

34 a The friend of Dave washed himself
b Dave's friend washed himself

children as young as three will successfully pick (*the*) *friend* in both cases, despite the fact that *Dave* is the nearer NP in (34a). As the structures below show, only the subject DP node (starred) c-commands the reflexive, and the referent of that DP is the entity *the friend*:

Structure for 34a Structure for 34b

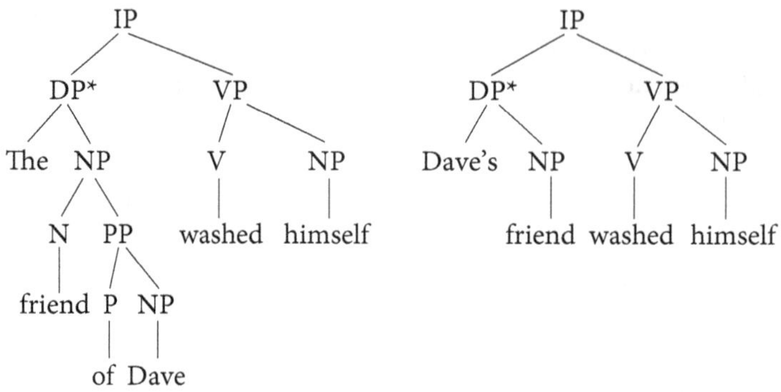


Studies which show sensitivity to c-command in selecting a referent within the local domain include Jakubowicz (1984), Deutch et al. (1986) and Chien and Wexler (1990).

However, although children are clearly sensitive to structural properties crucial to the operation of the binding theory, their performance on tests of knowledge of the principles is not always adult-like or at a high level. A persistent observation across many studies is that children do worse at interpreting definite pronouns than they do in interpreting reflexives. Why should that be?

One line of thinking is that the errors children commit in interpreting definite pronouns reflect muddle generated by the range of possible referents for pronouns (they may refer to an entity inside the sentence or to an entity present only in the discourse environment) and by the fact that there are circumstances in which the strictures of Principle B may be relaxed. For example, there are discourse-related circumstances in which the definite pronoun can be taken to refer inside the same clause. In the following exchange, Speaker B intends *him* and *John* to refer to the same person:

35 Speaker A: I don't know anyone who likes John
Speaker B: John likes him

If children do not fully command the use of discourse-related hedges on Principle B, then this perhaps is the source of their errors. Some support for this comes from the fact that cross-linguistic studies show that in languages in which the definite pronoun is combined with and precedes the verb by a process of cliticisation, as in the Italian example (36), the difference in performance on Principle B and Principle A disappears:

36 Lo gnomo lo lava
The gnome him washes

Examples of studies which show a high level of performance with clitic pronouns are McKee's (1992) study of Italian (from which example (36) is taken), and a study by Goodluck et al. (1995) of Serbian. Since clitic pronouns do not enter into the sort of discourse-related overrides of Principle B illustrated in (35), the rise in performance with Principle B when the pronoun is a clitic provides support for the view that children do know the binding theory, but have trouble executing it in circumstances where the grammar offers the possibility of relaxing its conditions.

A second line of thinking with respect to children's Principle B errors also holds that children do know the binding theory, but that the formulation of the theory sketched above is not quite right. Scholars such as Reinhart (1983) and Grodzinsky and Reinhart (1993) have suggested that the binding theory should regulate only cases of binding – linking an element to a c-commanding antecedent, as in the case of reflexives. Cases of co-reference which do not involve binding are held to be regulated by a different set of principles, that refer to whether or not there is an alternative way to express the intended proposition by using binding – if so, the co-reference is blocked. On this type of account, what children lack is skill in executing the rules that govern co-reference, not skill in executing the binding theory.

Some evidence which seemed to favour such an alternative account appeared to come from studies in which the potential referent of the pronoun was a quantified NP, such as *every X*. Chien and Wexler (1990) found that children's errors of incorrect co-reference between the pronoun and the subject dropped markedly in sentences such as (38) as compared to (37):

37 Is Goldilocks pointing at her?
38 Is every bear pointing at her?

The better performance on (38) is explained as follows: for both sentences, Principle B blocks the binding of *her* to the subject NP. Co-reference, determined at some level of pragmatic/discourse structure, is more readily permitted in the case of (37) than (38) because a name, such as *Goldilocks*, establishes a referent in the discourse representation that acts as a target for co-reference, whereas a quantified NP such as *every bear* does not establish such a target. Many, but not all, studies have since found better performance for quantified NPs.

In an extensive review of the literature and with original experiments, Conroy et al. (2009) challenge the results concerning (37) and (38), following concerns raised by Elbourne (2005). Their experiments used a Truth Value Judgement Task, similar to that used by Deen and Timyan (2018), in which the child was told a short story, after which s/he would be asked if a statement was correct or not. Their first experiment was critical in that the prominence of potential antecedents was deemed equal and the same story was used for both the quantified and non-quantified conditions. No difference between the conditions emerged. The conclusion that Conroy et al. draw is that the advantage that quantified antecedents have may be an artefact of the experiments in which the advantage was found.[6]

Nonetheless, in their review of the literature, Conroy et al. do find a small residue of an effect: it seems that pronouns are somewhat more difficult to process when the pronoun is subject to Principle B than when it is involved in a Principle C violation (as binder of an R-expression). Conroy et al. draw on the literature from adult sentence processing studies to explain this. In adult studies, it has been found that inappropriate antecedents that precede the pronoun are examined as well as legitimate antecedents, with the former being rejected. They argue that children are less able to complete the processing operation (with rejection of the illegitimate antecedent) than adults. The difference between pronouns subject to Principle B vs. pronouns involved in Principle C violations is that in the Principle B cases the pronoun follows its potential antecedent, enhancing the chance of errors for Principle B.

To summarise the results so far discussed with Principle B, it appears that children are aware of the dictates of the principle. The effect of using a clitic pronoun seems to be firm and argues for a core of knowledge when the potential for interference from discourse-related phenomena does not intervene. The apparent effect of quantified vs. non-quantified antecedents has been argued to result from flawed experimentation, again leaving the path clear for children's knowledge of the principle. Thus, the choice between Reinhart's/Grodzinsky and Reinhart's version of the binding theory and the 1981 version in Chomsky's work therefore cannot be made on the basis of these results.

However, furthering work by Reinhart and Reuland (1993), Principle B has been reformulated by Reuland (2001) in the context of more recent theory (the Minimalist Program). The effect of this reformulation is examined in child language by Ruigendijk et al. (2011), who also cite Philip and Coopmans (1996) and Baauw (2002). Reuland proposes that there is a difference between sentences of type (39) and sentences of type (40). In both cases, co-reference between the subject noun phrase and the pronoun is ungrammatical in the adult grammar:

39 *De $jongen_i$ heeft $hem_i$ aangeraakt
The boy has him touched
'The boy touched him'

40 *De $jongen_i$ zag $hem_i$ dansen
The boy saw him dance
'The boy saw him dance'

In the binding theory of Chomsky (1981), both are treated as Principle B violations. The local domain in (40), which contains two verbs, is the whole sentence, as it is in (39); the higher verb in (40) is one of the verbs in Dutch (and English) that case mark as accusative the subject of the lower verb (*hem*), and this determines the boundary of the domain as the whole sentence.

But the source of the ungrammaticality of (39) and (40) is different in Reuland's analysis. In (39), the ungrammaticality is ascribed to the violation of a condition requiring that reflexive marking is required for a reflexive interpretation. Example (40) is ungrammatical not by virtue of this condition, but because it violates a restriction relating to discourse co-reference.

How do children treat (39) vs. (40)? Ruigendijk et al. (2011) find that children make many more errors with (40) than with (39), which they interpret as evidence in favour of Reuland's theory of anaphora.[7] Thus, this example is one in which one theory of grammar makes a prediction but another theory (the Government and Binding theory of Chomsky) does not.

*Are children always accurate with Principle A?* Although children do very well at a young age with straightforward cases of the interpretation of reflexives in studies such as those cited above, there are more complex systems that may take years to master. An example is the acquisition of reflexives in Danish in a study by Jakubowicz (1994). There are two reflexive pronouns in Danish, one complex (*sig selv*, equivalent to English *himself/herself*) and the other simple (*sig*). The complex reflexive acts syntactically in the same way as the English reflexive, requiring an antecedent inside the local domain. The simple reflexive by contrast requires an antecedent *outside* the local domain, when the verb accompanying the reflexive is one of a certain semantic class, called 'nonaffectedness' verbs. The difference between affectedness and nonaffectedness verbs is broadly whether a reflexive action involves physical contact (affectedness verbs) or not (nonaffectedness verbs). The verb meaning 'point at' is a nonaffectedness verb and requires a non-local antecedent with the reflexive *sig*:[8]

41 a Minnie$_i$ beder Ida$_j$ om at page pa sig selv$_{*i/yesj}$
Minnie asks Ida Comp point at REFL SELF
'Minnie asks Ida to point at herself (Ida)'
b Minnie$_i$ beder Ida$_j$ om at page pa sig$_{yesi/*j}$
Minnie asks Ida Comp point at REFL
'Minnie asks Ida to point at herself (Minnie)'

Although children in Jakubowicz's study did very well with the complex reflexive from an early age, at age nine they were still making a substantial proportion of errors in interpreting the simple reflexive in sentences that required an antecedent outside the clause.

### 4.4.2 THE DEVELOPMENT OF MOVEMENT

Movement is both local (linking the subject of passive sentences to the object position) and long-distance (linking the position of question words and other phrases to their base position over stretches of potentially infinite length).

*Passive.* In sentences such as (42), the subject is understood as the logical object of the verb:

42 Denise was arrested (by the police) for her unruly behaviour

Example (42) is formed from a structure in which the object moves into an empty subject position:

43 [e] was arrested Denise (by the police) for her unruly behaviour

The parentheses around *by the police* indicate that the agent phrase is optional.

An additional complexity in English derives from the fact that there are two types of passive, one formed by movement (as in (42)) and the other formed without movement – the adjectival passive. In (44):

44 The ice was broken

it is hypothesised that there is no movement operation. Adjectival passives resist the presence of a *by* phrase and express a state of being. In (44) we can prefix the verb with *un-*, making a *by* phrase with an agentive reading impossible:

45 The ice was unbroken by Harry

In (45) the *by* phrase has only a location interpretation (*The ice was unbroken near Harry*), not an agentive interpretation.

Passive sentences show a developmental delay in English, as compared with active sentences, and this delay has been linked to the verb that is passivised. For example, Maratsos et al. (1985) found that the full passive (with a *by* phrase) of actional verbs such as *wash*, in which the subject is an agent in the active, was comprehended better than the full passive of verbs in which the subject is not an agent in the active. That is, a passive such as (46) was better comprehended than a passive such as (47):

46 The bear was washed by the fox
47 The bear was seen by the fox

Such results led Borer and Wexler (1987, 1992) to propose that, for children, all passives are adjectival, with no movement involved. Borer and Wexler sketch restrictions on sentences that apply to non-agentive verbs,[9] arguing that such restrictions entail that an adjectival passive is possible only for verbs that have an agentive subject.

A number of studies have challenged the hypothesis that children's passives are formed without movement, at least in some languages. Demuth et al. (2010) investigated the passive in children learning Sesotho, a Bantu language spoken in South Africa. The Sesotho passive occurs with a higher frequency than the English passive, and the passive also occurs with the equivalent of a *by* phrase more frequently (60 per cent in Sesotho vs. 4 per cent in English). Adjectival constructions have a quite distinct morphological marking in Sesotho, unlike in English, in which the morphology is the same (the past participle is used in both examples (42) and (44)) whether movement is involved or not. Three-year-old children learning Sesotho showed comprehension and production of passive sentences far superior to their age-equivalents learning English.

In the acquisition of English, Fox and Grodzinsky (1998) found that performance of three- to five-year-old children with non-agentive passives improved when the *by* phrase was omitted. Fox and Godzinsky proposed that the operation of movement in the passive sentence was not impaired, but rather the problem with non-agentive verbs was with transmitting a non-agentive theta role through the *by* phrase.[10] A recent study by Nguyen and Pearl (2017) reviews the literature on the passive in English, concluding that it is the lexical semantic features of the passivised verb that determine the success in comprehending the passive, as opposed to an individual verb's frequency, or frequency in passive syntax.

Complementing such studies as Maratsos et al. and Nguyen and Pearl, researchers such as Gehrke and Grillo (2009) have suggested an approach to the passive in which passives are formed by an operation on event structure, syntactically realised (a concept not dealt with here). They exploit both old and recent theory to propose that passives involve a topicalisation of a subpart of a complex event. The details of this proposal are beyond our scope in this chapter, but it accounts for a range of phenomena that have been difficult to analyse on an approach simply based on movement of the object NP, including the fact that certain verbs do not passivise (the active *This laptop weighed two kilos* cannot be converted into the passive **Two kilos were weighed (by this laptop)*).

*Long-distance movement.* In questions and other sentence types which deviate from the normal subject-verb-object word order in English, movement is used to achieve that order. A question may be formed from a single clause, or it may be formed by movement over a potentially unbounded distance:

48 Who kissed Sue?
49 Who did Sue kiss?
50 Who did Fred think that Sue kissed?
51 Who did Fred imagine that Bill thought that ... Sue kissed?

Despite the potential to create an infinite number of questions, there are structural limits on question formation. There are 'islands' into which a question word cannot penetrate. We cannot in English question from within a relative clause (52), from within an embedded question (53), or from within an adverbial clause (54):

52 *Who did Fred see a man that kissed?
53 a *Who did Fred wonder who kissed?
b *When did Fred say how he hurt himself?
54 *Who did Fred kiss Sue before he danced with?

In (53a) an object NP (*who*) has been moved out an embedded question, and in (53b) an adjunct phrase (*when*) has been moved out of an embedded question (the question is grammatical on the reading in which *when* modifies *say*, but ungrammatical if *when* modifies *hurt*).

It might be thought that questions such as (52–54) are too complex cognitively – that they don't 'make sense'. Such a possibility is ruled out by the existence of languages that permit the equivalent of (52–54) to some degree. An extreme case is Akan, a Kwa language spoken in Ghana, in which all the equivalents of (52–54) are grammatical. (For Akan, a different derivation has been proposed, in which questions are formed by pronominal binding of the question word to the position it occupies in underlying structure; see Chapter 7, section 7.3.3.)

Children's production of wh-questions with *what* and *where* emerges in Table 4.2 at around twenty months. Recent experimental work has deepened our knowledge of children's ability, with results suggesting that question formation may be in the grasp of infants as young as fifteen months. Seidl et al. (2003) used a technique in which the infant's gaze time was recorded. The infants saw videos in which, for example, a book hits a set of keys, or the reverse action in which the keys hit the book, followed by a split screen in which the book and the keys appeared. The videos were appropriate for either a subject question (55a) or an object question (55b):

55 a What hit the keys?
   b What did the keys hit?

The time that the infants spent looking at the objects on the split screen was measured. At fifteen months and at twenty months infants looked longer at the book as opposed to the keys for subject questions (55a), and at twenty months they also looked longer at the book than the keys for object questions (55b). This comprehension result lowers the age at which production of *what* questions is reported (Table 4.2).

A study by Stromswold (1995) also investigated subject and object questions in corpora from the CHILDES database. The data from twelve children aged between 1;4 and 6;0 were examined. Stromswold found that children acquire object questions before subject questions, somewhat to the contrary of the study by Seidl et al. (the apparent disparity may be resolved in terms of the number of subjects and/or ages tested).

Are children sensitive to the structural constraints on question formation, illustrated by (52–54)? Otsu (1981) and de Villiers et al. (1990) were the first to tackle this question. The results of Otsu were suggestive of knowledge of the block on extraction from a relative clause (52) and those of de Villiers et al. were suggestive of knowledge of the block on extraction from an embedded question in (53a–b). For example, de Villiers et al. used a task in which three- to six-year-old children answered questions following short stories accompanied by pictures that potentially provided two answers, one of which violated the constraint. An example of such a story and question combination is given in (56):[11]

56 Story: This boy loved climbing trees.
One afternoon he climbed this tree, but
look, he slipped and fell.
He picked himself up and went home.
That night when he had a bath, he found a big bruise on his arm.
He said to his dad: 'I must have hurt myself when I fell this afternoon.'
Question: When did the boy say how he hurt himself?

If the child follows the block on movement from within an embedded question, the answer should be *that night* or *in the bath*, not *that afternoon*. And, indeed, there were significantly more responses with movement out of the top clause than there were with movement out of the embedded question. Moreover, it was not simply a preference for linking a question phrase with the main clause, as children gave lower clause extraction answers (*that afternoon*) to the question:

57 When did the boy say he hurt himself?

However, the findings were not all that clear-cut (see Weinberg 1990 for one discussion), although the basic observation is replicable (de Villiers et al. 2008). Later data on the block on extraction from an adjunct clause (54) and from a relative clause (53) appeared to be more categorial (Goodluck et al. 1992; de Villiers and Roeper 1995). But the evidence for these blocks on extraction may be confounded by processing preferences (see Chapter 7 and Goodluck 2007).

### 4.4.3 CHILDREN'S GRAMMAR OF CONTROL

The grammar of sentences with missing subjects is one of the thorniest topics in syntactic theory (see Landau and Thornton 2011 for representative references). As we saw in section 4.1, the missing subject of tenseless subordinate clauses is represented by an unpronounced element, PRO. Infinitival complements to verbs such as *want*, *tell* and *choose* (58a–c), tenseless temporal clauses (58d) and infinitival subject clauses (58e) are among the clause types that have a PRO subject:

58 a Fred wanted [PRO to leave]
 b Fred told Jane [PRO to leave]
 c Fred chose Jane [PRO to leave]
 d Fred kissed Jane [before PRO leaving]
 e [PRO to kiss Jane] would be a crime

Despite the complexity of the grammar of PRO, there are some generalisations that can provide a framework for evaluation of children's knowledge. First, as mentioned in section 4.1, control may be either obligatory or optional. In the case of (58a–d), the PRO must be interpreted as referring to another NP in the sentence. In the case of (58a), there is only one NP in the sentence. In the case of (58b–c), the controller (the NP interpreted as referring to PRO) is the main clause object (*Jane* in the examples). In the case of (58d), the main clause subject (*Fred*) is controller. For these clause types, PRO is obligatorily interpreted as co-referent with an NP inside the sentence. In contrast, the PRO in (58e) must be interpreted as referring to an entity not mentioned in the sentence, although sentence-internal reference is possible (in *To kiss Jane would please Bill*, Bill may be the one who does the kissing). Control is thus optional.

A number of properties have been proposed as characteristic of obligatory control constructions, one of which is whether or not there is a c-command relation between PRO and the NP it refers to. In (58a–c), where the subordinate clause is attached to the VP node, both the main clause subject and object (if there is one) c-command the PRO; in (58d), with a clause that attaches to the main clause IP node, only the subject c-commands PRO; in the structure (58e), there is no NP inside the sentence that c-commands PRO:

Structure for 58a

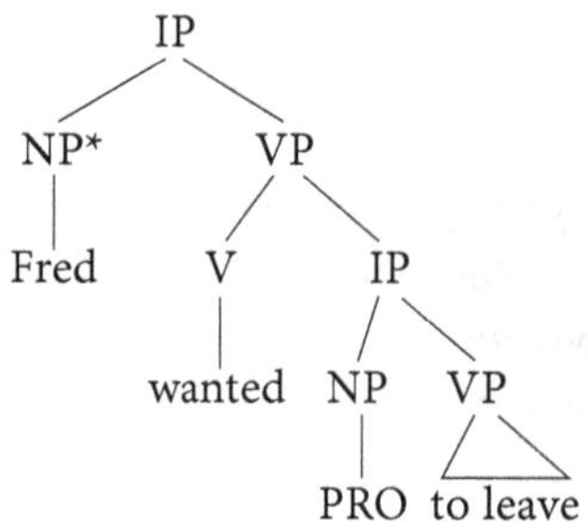


Structure for 58d

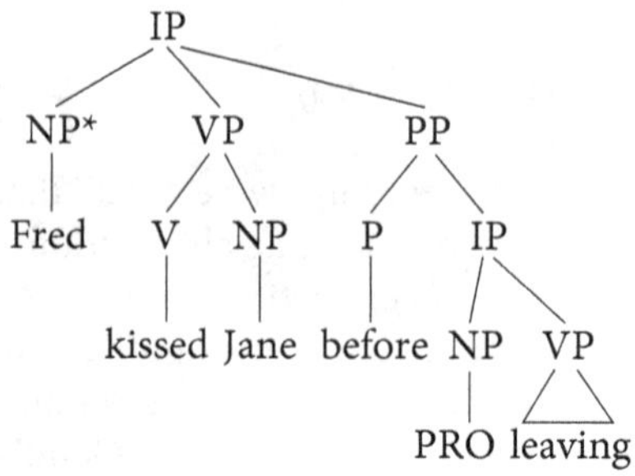


Structure for 58b–c

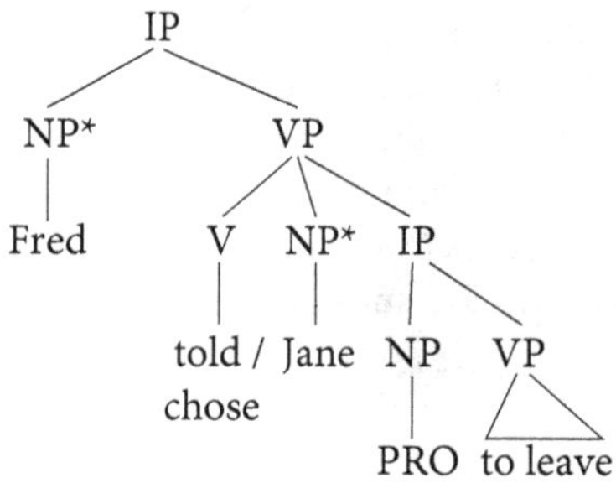


Structure for 58e

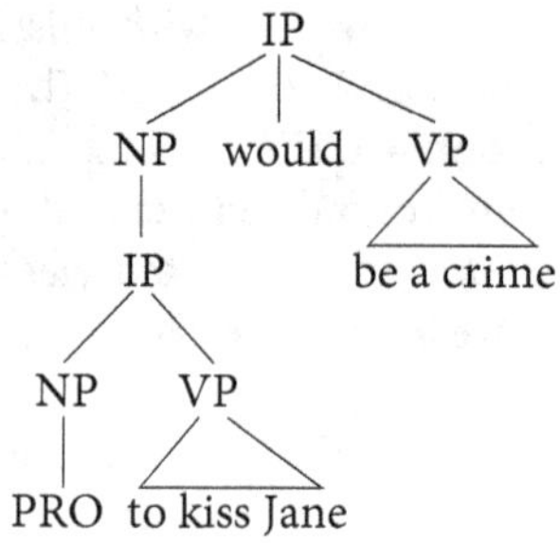


(NPs that c-command PRO are starred.)

Second, control for complements in the VP (such as 58b–c) is semantically (thematically) determined. For such complements, the general consensus is that control rules make reference to the thematic structure of the main clause, and that this requires some relaxation of the c-command condition.[12] In examples such as (58b–c), the controller is the NP (*Jane*) with the thematic role of *theme* or *goal*. This is true for a good many verbs in English. The verb *promise* breaks with this pattern, with the subject NP as controller:

59 Fred promised Jane [PRO to leave]

How do children handle PRO constructions? Recent work on the theory of control and on child language has focused on the very early stages, with children as young as eighteen months. Adult English makes only a very restrictive use of *subjunctive* complements; subjunctive complements do not impose agreement on the subject of the complement and the verb in the complement, such as the embedded clause in (60) with the infinitive *be*:

60 The administration ordered that the ceremony be cancelled

Other languages such as Greek, however, use the subjunctive much more widely, and they use it to express both optional and obligatory control relations.[13] One hypothesis concerning children's early grammar of control in English is that children impose a subjunctive structure on strings such as the complement to *want*.[14] Evidence in favour of this is presented by Landau and Thornton (2011). In a study of one child, Laura, aged one to two years, Landau and Thornton found that she used the main verb *want* in contexts such as the following:

61 a Context: Laura wanted mother to push her in the stroller
Laura: I want _ push Laura
b Context: Laura wanted mother to dance
Laura: I want _ dance

The context makes it clear that someone other than the main clause subject *I* is intended as subject of the subordinate verb. Evidence in favour of an interpretation of the complement as a subjunctive comes from the fact that almost all (7/8) of the subordinate verbs lack agreement in those third person cases where agreement is visible in English, contrary to the child's use of agreement in other constructions. Thus, the pattern of this child's performance argues in favour of the child plucking an analysis from the tools made available by Universal Grammar, although not used in her ambient language. By the time Laura was approximately twenty-eight months, she had introduced *to* into her productions of the verb *want* and dropped the option of reference outside the sentence.

Many studies have focused on children's somewhat later performance, from age three to six. The main findings for obligatory control can be summarised as follows:

- Object control of the complement to verbs such as *tell* and *ask* is accurate from the earliest ages studied, and, moreover, control by the surface subject of passive sentences with *tell* and *ask* is also accurate (in *John was told by Bill to leave*, *John* is the one who leaves), provided the passive is correctly understood (Goodluck 1981; Hsu et al. 1985; Goodluck and Behne 1992; Janke 2018a).
- Object control is overgeneralised to the verb *promise* until age five or six (Chomsky 1969; Maratsos 1974b; Goodluck 1981; but see Cohen-Sherman and Lust 1993 for contrary argument, and Cairns et al. 1993 for discussion).

The results for optional control, which is reliant on pragmatic influences, are taken up briefly in Chapter 7 (p. 125).

## 4.5 LATER DEVELOPMENTS

We have already seen one case in which children take time to master the link between lexical items and grammar: children in Jakubowicz's study were as old as nine before they grasped the need to refer outside the clause for the reflexive *sig* when the verb belongs to the 'nonaffectedness' class. Most cases of putative late learning involve the connection between lexical items and rules. However, we will begin with a case in which there is no such restriction: the control of adjunct clauses, which we have seen is determined by the only c-commanding entity, the subject (see 58d), not by the properties of the main clause predicate.

Children are typically aged seven or older before they reliably follow the rule for (58d). Before that, children may make a number of errors, although the response pattern in any given study is rarely random. They may treat the PRO subject in the adult grammar as arbitrary in reference, or they may opt for control by a thematically determined NP (either agent or patient); see Goodluck (1981), Hsu et al. (1985), McDaniel et al. (1990/1991) and Goodluck and Behne (1992). Goodluck (2001) argued that the correct analysis was that children impose a nominal analysis on the adjunct clause, as proposed by Carlson (1990) and adopted by Wexler (1992), i.e. they construe a sentence such as:

62 Snowy pushes Leo before [$_{IP}$ PRO dancing by Ellie]

as having the approximate structure:

63 Snowy pushes Leo before [$_{NP}$ dancing by Ellie]

A nominal structure allows essentially free reference of the subject of the subordinated structure, whereas the correct interpretation of PRO involves restrictions. In particular, it requires that the phrase *by Ellie* is given a locative interpretation (the dancing takes place near *Ellie*), not an agentive interpretation (*Ellie* dances). The essence of the block on an agentive argument for *Ellie* with a PRO structure is the fact that only one agent per clause can be assigned, and if a PRO structure is used, then the agent role must go to the PRO. Asked to act out sentences such as (62), children aged four to six freely gave an agentive interpretation to the *by* phrase, while adults never did. This supports the nominal analysis of children's interpretations of adjunct PRO constructions.[15]

The development of constructions such as (9) (*Children are nice to understand*) has been the subject of many studies (see Becker 2014 for a review). Most adjectives fall into one class or another – i.e. those adjectives that have a controlled PRO in the embedded clause, such as *eager*:

64 Jane is eager [PRO to kiss]

and those adjectives such as *easy* that contain a missing object in the embedded clause, with the subject of that clause arbitrarily interpreted:

65 Jane is easy [PRO$_{ARB}$ to kiss [e]]

The earliest studies, in which children acted out their interpretations, suggested that the more complex structure (exemplified by *easy*) was a late acquisition; until seven to eight or more years of age, *easy* adjectives were interpreted as *eager* adjectives, with the main clause subject co-referential with the embedded subject (Chomsky 1969;

Cromer 1970, 1987). Anderson (2005) used a Sentence Judgement Task and found inconsistent use of the interpretation appropriate for *easy*-type adjectives, suggesting that that interpretation is available to children younger than six. Becker et al. (2012) and Becker (2014) report an experiment in which nonsense adjectives were taught to children aged four to seven years. The children were divided into two groups, one of which was presented with sentences with nonsense adjectives that had an inanimate subject (66) and the other with sentences that had an animate subject (67):

66 Apples are very daxy to draw
67 The policeman is daxy to draw

Video presentations of scenarios appropriate to an *easy*-type interpretation of the nonsense adjective were presented, and reaction times to answering a Yes-No question following the video were measured. Reaction times to the question:

68 Is it daxy to draw apples?

were shorter for children who heard sentences with inanimate subjects than were reaction times for the question:

69 Is it daxy to draw the policeman?

for those children who heard sentences with animate subjects. Longer reaction times are associated with ungrammaticality/difficulty. Becker (2014) argues that (in)-animacy provides an essential clue to the existence of a displaced object, a clue that may be used for several different structures (see section 4.7 below).

## 4.6 SYNTACTIC DEVELOPMENT AND THE RECENT THEORY OF GRAMMAR

The previous sections have almost completely avoided mention of developments in Chomskyan syntactic theory that have been ongoing since the 1990s. The 1981 model of grammar – the Government and Binding theory – contained distinct levels of representation: D-structure and S-structure (deep structure and surface structure). D-Structure was created by phrase structure strictures and the lexicon. The levels of D-structure and S-structure were linked by the operation of movement, and movement applied again to produce a third level, Logical Form (LF), which formed the basis for our understanding of sentences. This model of the organisation of grammar is represented in Figure 4.1. Since 1993, the Minimalist Program has developed a model of the organisation of grammar in which D-Structure and S-Structure no longer exist as distinct levels of representation, a model that looks like that presented in Figure 4.2. In the model in Figure 4.2, a sentence, or a fragment of the sentence, is pared off to what is termed the conceptual interface when it is 'ready' – i.e. when the requirements of the syntax have been met. See Chomsky (1993) for a first exposition of the goals of the Minimalist Program. Edwin Williams memorably described

Phrase structure ----- D-STRUCTURE ----- Lexicon
|
Movement
|
S-STRUCTURE
|
Movement and rules of interpretation
|
LOGICAL FORM

*Figure 4.1 The Government-Binding Model*

Lexicon
|
Numeration
|
Derivations governed by principles of the binding theory, the bounding theory and control theory, *inter alia*

| | |
|---|---|
| Logical Form | Phonetic Form |
| Conceptual-Intentional Interface | Articulatory-Perceptual Interface |

*Figure 4.2 The Minimalist Model*

the difference between the model of grammar in *Aspects of the Theory of Syntax* and the model in *Lectures on Government and Binding* as a difference between an essay question and a multiple-choice quiz. We might characterise the difference between Government and Binding theory and the Minimalist Program as the difference between a three-act play and a series of skits of variable length.

It is beyond the scope of this chapter to consider the motivation for the changes between Government and Binding theory and the Minimalist Program. But we can at least consider some ideas about how well the new model does in accounting for children's development.

In the Minimalist Program, words are inserted at the beginning of the derivation – at a representation called the numeration – fully inflected. The phrase structure is built up piece by piece using a binary operation Merge, which combines two words or two phrasal units together. It is tempting to analyse the speech of the children in Table 4.1 – which shows a clear 'two-word stage', particularly in the case of Gregory – as reflecting the first attempts at Merge. And the utterance sequence of Abigail in (23) shows that although she is capable of outputting several cases of Merge in a single utterance, she still plays with the small combinations; her fragments in (23) may represent some of the sequences along the path towards the final utterance. (A fuller account of the development of Merge is given in Goodluck and Kazanina 2020.)

The lexicon in recent theory also contains syntactic features that determine the derivation that follows in the syntax. In Reuland's (2001) analysis, for example, verbs are specified as plus or minus 'reflexive', and this determines whether the verb is accompanied by a complex reflexive pronoun or a simple reflexive pronoun, whose features in turn require validation ('checking') in the derivation that follows. And we have seen the idea of incorporation as an explanation of the difference in grammaticality between regular and irregular nouns in compounds, the former requiring syntactic features to be checked, and the latter not (Chapter 3).

An overall coherent narrative of the development of such syntactic features remains a goal that has not yet been a major focus of acquisition studies. It is possible that the meaning of verbs in and of itself may drive the child towards an analysis. On the other hand, the distinction between +/– affectedness verbs in Danish – requiring reference outside the clause in which the verb is embedded for the simple *sig* reflexive if the clause is headed by a -affectedness verb – appears to be a late development. Whether it is knowledge of the lexical distinction between +/– affectedness verbs, or a processing preference for a local antecedent, is not known at present.

Another issue that has come to the fore in recent theory concerns the distinction between competence and performance. We presented this in Chapter 1 as a clear-cut division between what the native speaker knows and what s/he may do in actual speech acts. But many recent theoretical analyses in effect blur the distinction by building in sequences of operations that rely on the need to access memory resources and/or the need to access broader domains of discourse. An example again is Reuland's (2001) work: Reuland proposes that derivations which require access to discourse representations more costly than those which do not. Such intertwining of pure theory with performance mechanisms may have the desirable result of explaining cases in which adult judgements are other than clear-cut, and of explaining children's behaviour (as in the case of Ruigendijk et al.'s 2011 study).

Finally, recall that in Chapter 3 we left largely unexplained the difference between English, in which the impoverished morphology may take years to master, and languages such as Inuktitut, in which a complex yet 'transparent' morphological system seems to be acquired with great ease. Perhaps the idea that lexical items are entered into the derivation fully inflected may yield some insight into this difference; as we will see in the next chapter, English requires the shifting around of inflectional elements to complete the derivation, operations that are potentially avoided in languages such as Inuktitut.

## 4.7 SEMANTIC BOOTSTRAPPING, PROSODIC BOOTSTRAPPING AND BEYOND

Are there principles or processes which the child uses which tap into regularities in the language s/he is learning and languages in general? How does the infant make sense of the speech stream and crack into the underlying linguistic system? A recurring idea is that the child is equipped with knowledge of the probability of a particular syntactic category mapping into a type of entity in the world, and that, combined with sensitivity to phonetic and prosodic features of the speech s/he hears, allows the child to construct a primitive syntactic system.

Semantic bootstrapping exploits the fact that there are regularities in both the assignment of syntactic categories and in the distribution of thematic roles (agent, patient/theme, etc.). The category Noun, for instance, is prototypically the name of a person or a thing and Verbs prototypically express an action or a change of state. Agents are most often subjects, patients are most often objects, and goals and locations are frequently realised in prepositional phrases and/or with particular morphological marking. Pinker (1984, 1987) proposes that these regularities bootstrap the child into the formation of syntactic rules using a set of primitive categories (Noun, Verb, Adjective, Preposition); similar ideas were sketched in Grimshaw (1981). That is, the child at the very earliest stages assigns word categories according to these prototypical values, and then projects a syntax (a noun phrase subject precedes a verb phrase) that exploits generalisations concerning the mapping between thematic roles and syntactic categories.

But how does the child work out what a word is in the first place? A great deal of research has shown that phonetic/prosodic cues are available for word boundaries, and that very young infants are tuned to the phonetic/prosodic structure of their language. For example, a constellation of acoustic cues signal the difference between lexical, open-class words (nouns, verbs, adjectives and adverbs) and functional, closed-class words (prepositions, articles, etc.). Closed-class words tend to have shorter vowel duration and have simpler syllable structure. Gow and Gordon (1995) and Gow et al. (n.d.) have argued that the onsets of words have distinct properties (for example, voice onset time (Chapter 2) is a reliable cue for syllable initial consonants but not syllable final consonants).

In sucking experiments of the type described in Chapter 2, Shi et al. (1999) found that newborn (1–3-day-old) infants could discriminate between lexical words (nouns, verbs, adjectives and adverbs) and function words (prepositions, articles and pronouns), with an increased sucking rate when two lists were presented that changed from one type of word (lexical or functional) in the first list to another type of word in the second list, as opposed to two lists with the same type of word, for which no significant increase in sucking rate occurred. In a head turn experiment, Johnson (2008) found that twelve-month-old infants looked longer at words that obeyed the prosodic structure that they had been trained on – that is, they looked longer at a word (e.g. *toga*) if they had been trained on passages that contained repeated instances of a stress initial bisyllabic word followed by a monosyllabic word (*toga#lore*) than if they had been trained on repeated instances of a monosyllabic word followed by a stress final bisyllabic word (e.g. *toe#galore*). Such results indicate that word boundaries are phonetically distinct, and that infants can use phonetic cues to locate such boundaries.

Semantic bootstrapping and phonetic/prosodic bootstrapping address questions about the very earliest stages of linguistic development. Grammar development takes place based on probabilistic relations between the structure of languages in general and the child's hypotheses – the child is unconsciously aware that subjects are likely to be agents. Such bootstrapping mechanisms potentially allow the foundations of the linguistic system to be built, subject to UG.[16]

Is there any role for such probabilistic projections of knowledge in later development? The work by Becker (2014) reported above suggests there is, and that the effects extend beyond *easy*-type adjectives. Movement can involve not only objects (as in the case of *easy* predicates), but also subjects. In (70), the subordinate subject has been moved from the embedded clause to occupy the main clause subject position:

70 Bert seems to Ernie [e to be wearing a hat]

Wexler (2004) and Hirsch et al. (2007) document children's difficulty with sentences in which the lower subject has been raised to the higher subject position, and Becker (2014: 194) found only three instances of such constructions in almost a quarter of a million spontaneous utterances by children (the three utterances were from children aged three and a half and almost five). Again, Becker (2014) reports improved performance for children aged three to five when the raised subject is inanimate.

## CHAPTER SUMMARY

Children make rapid progress in the acquisition of syntax between the ages of two and three. The so-called period of telegraphic speech gives way by age three to an array of sentence types that includes pretty much all of the sentence types in the adult grammar. Although at age three children do not have a fully adult grammar of the language they are exposed to, an increasing body of data argues that they are sensitive to restrictions evidenced in adult grammars – for example, children display limits on their production of subjectless sentences similar to those found in adult English dialects. Of the period between three and five to six years, the binding theory and the development of movement in passive sentences have received the most attention. The principles of the binding theory are present in children's language, and the evidence for these principles is of course evidence for the structures upon which the binding theory is built. Movement in passive sentences is more controversial, with data showing that such movement may not emerge until four to five years (the child may rely on an analysis that does not involve movement, but their analysis is one that is present in adult grammars). The evidence for restrictions on long-distance movement (island constraints) is more problematic, and we return to this topic in a later chapter (Chapter 7). Control theory (the interpretation of PRO) is developed at an early age for basic cases, such as the interpretation of PRO in active and passive sentences with main verbs such as *tell*. The acquisition of cases such as *promise* may take until around six years to master, and the interpretation of PRO in adjunct clauses is a later development still. We ended the chapter with a look at what drives the development of syntax: the semantic and phonetic/prosodic bootstrapping hypotheses, and recent research into the role of animacy in prompting knowledge of later acquired rules.

## FURTHER READING

Cook and Newson (2007) and Gallego (2011) are clear and informative guides to recent Chomskyan theory. The brave might tackle some of Chomsky's own writings (Chomsky 1993, 2005). Guasti (2016) is a text which covers many of the same topics in this chapter and Chapter 5, often in more detail; her views are not always the same as mine. Pérez-Leroux et al. (2017) is a recent book on the acquistion of direct objects, a topic not included in this chapter.

## QUESTIONS AND EXERCISES

1. Solan (1983) reports the following percentages of co-reference between the pronoun and the NP *the sheep* for five-year-olds for sentences (i–iv):

   | | |
   |---|---|
   | i. He told the horse that the sheep would run around | 17% |
   | ii. The horse told him that the sheep would run around | 17% |
   | iii. He hit the horse after the sheep ran around | 11% |
   | iv. The horse hit him after the sheep ran around | 39% |

   What is the difference in structure between (i–ii) and (iii–iv) (see the trees in 58b–d)? How does this support c-command as the restriction on pronominal co-reference?

2. Compare the results of Landau and Thornton (2011) with Goodluck (2001). Both studies suggest that control is free, in the case of Landau and Thornton at the earliest stage of control of complements to *want* and in the case of Goodluck for control of adjunct clauses through the kindergarten years. What might cause very rapid development in the one case, but not the other? Can you propose an alternative that will cover both sets of data?

3. De Villiers et al. (1990) observe that the earliest examples of question formation from an embedded clause in the speech of Adam aged three and a half (Brown 1973) used the question word *what*:

   i. What chu like to have?
   ii. What (d')you think this look like?
   iii. What he want to play with?

   Can you connect this to any other findings reported in this chapter about the development of grammar?

4. Find a language or languages other than English on the CHILDES database. For any given phenomenon described in this chapter (for example, the existence of subjectless sentences or the order of emergence of embedded clauses), check the facts against the language(s). Important points to remember: (1) find out the facts in the adult grammar in your chosen language(s) before you begin; (2) age is only

a rough guide to appearance in a corpus. This exercise is probably best tackled by a number of students working together as a group.

## NOTES

1. IP has replaced S(entence) in most recent syntactic accounts; IP is an abbreviation for a number of types of phrases (see Chapter 5).
2. The structure is still possibly not accurate. In some analyses, the main clause subject is not moved from object position of the embedded clause but is linked to an abstract element O (operator) that moves from the object position to the front of the embedded clause.
3. There are exceptions to this generalisation; we will see in Chapter 4 that morphemes associated with tense and aspect may move down to an adjacent element.
4. The notation *( ) indicates that omission of the element inside the parentheses is ungrammatical.
5. In the literature on the development of the binding theory, it has been stated that Lust and/or Solan did not pay attention to structural constraints but only to a linear restriction, blocking co-reference between a pronoun and an NP that follows it. This is incorrect. For example, Solan discusses the plausibility of a linear restriction (concluding on the basis of cross-linguistic evidence that such a restriction is plausible) and his data support the idea that c-command is the first alternative hypothesis against which a linear restriction – if it is present – is relaxed.
6. Chien and Wexler used a picture verification task; Conroy et al. argued that this is subject to similar restrictions on use as the Truth Value Judgement Task.
7. Ruigendijk et al. do not claim that children lack knowledge of the grammar governing (40), rather they propose that their ability to utilise the grammar fails when it requires access to discourse conditions – a performance limitation.
8. In a more articulated structure than that given in (41), *Ida* controls the PRO subject of the embedded clause:
   ... $\text{Ida}_i$ om at [$\text{PRO}_i$ page ...]
9. For example, non-agentive verbs are ungrammatical in sentences such as **The doll appears seen*, whereas agentive verbs are grammatical, *The doll appears combed.*
10. Hirsch and Wexler (n.d.) query whether Fox and Grodzinsky's result replicates. In unpublished work, Goodluck, Eriks-Brophy and Stojanović find the same result as Fox and Grodzinsky for a subset of typically developing children and some persons with Down syndrome.
11. Thanks to Jill de Villiers, who supplied this example, which is not included in de Villiers et al.'s article.
12. For example, *John* is controller in a sentence such as *Bill shouted to John* [PRO *to call the cops*], although as the object of a preposition it does not c-command PRO.
13. Accurate understanding of the subjunctive in control in languages such as Greek may take into the school years to develop (see Goodluck et al. 2001).
14. In Table 4.2, child utterances are reported using the form *wanna*. Such utterances were excluded from the analysis of Landau and Thornton (and other researchers) since their status is unclear. *Wanna* is not an earlier occurrence than *want.*
15. Earlier analyses attributed the errors made on adjunct clauses to misattachment of the adjunct clause to the VP (Goodluck 1981; Hsu et al. 1985), and misattachment of adjunct clauses has also been proposed in recent work by Janke (2018a). The misattachment hypothesis runs into the problem that results of Solan (1983) and Goodluck and Solan (2001) on pronominal reference (see p. 58) argue that correct attachment is known to the child.
16. Syntactic bootstrapping is a further proposed mechanism that allows the child to exploit the structure of the sentence to infer the meaning of lexical items; see Chapter 6, section 6.8.3.

CHAPTER 5

# FURTHER ASPECTS OF SYNTACTIC AND SEMANTIC DEVELOPMENT

In this chapter, we look at the child's knowledge of tense and aspect, of quantification, and of other lexical and syntactic distinctions. First, we review what is meant by tense and aspect, and how they are syntactically realised. We briefly summarise the literature on a debate concerning whether children's very early grammars are confined to only lexical (VP, NP, PP and AP) phrases, and on whether tense or aspect has priority in child language. Next, we examine negation and modality in child language. We then look at children's success and errors with quantification, and at some of the research that has challenged the position that children do not know how quantifiers work. Finally, we tackle children's over- and under-extensions of word meanings and their comprehension of count vs. mass nouns.

## 5.1 TENSE

Tense is realised in English in terms of plus and minus past (1a and b). Future is indicated by use of the present tense, with a modal verb *will* (1c, or a present tense periphrastic construction *is going to/gonna*),

1 a Harriet is happy (present)
  b Harriet was happy (past)
  c Harriet will be happy (future)

Adverbs can convey time, as indicated by the addition of *now* to (1a), *yesterday* to (1b) and *tomorrow* to (1c). Mixing tense with the wrong adverb yields ungrammatical/infelicitous results (as in **Harriet is happy yesterday/tomorrow*, **Harriet was happy tomorrow* and **Harriet will be happy yesterday*). Tense is realised on the first element in a sequence of auxiliary and main verbs, as shown in (2):

2 a Harriet is running
  b Harriet was running

### 5.1.1 THE SYNTACTIC SUPERSTRUCTURE FOR TENSE AND AGREEMENT

In syntactic theory, the node IP (Inflectional Phrase) is no longer used, except as an abbreviatory device. Rather, the IP node is fractioned into a cluster of functional phrases, to which belong AgrP (Agreement Phrase), TP (Tense Phrase) and AspP (Aspect Phrase). These functional phrases are distinct from lexical phrases such as NP and VP in that there is no word that projects up to head the phrase. Sentence (2a) will have the approximate structure in (3):

3

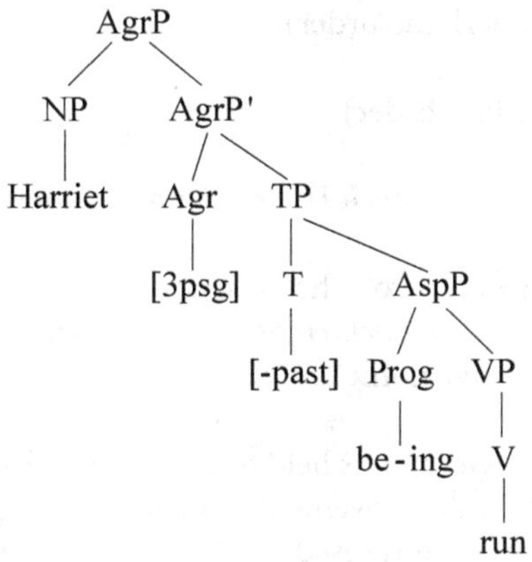


In (3), we see that the [-past] and [3psg] (third person singular) end up on the next lower lexical item (*be*), and the progressive *-ing* affix ends up on the verb below, *run*.[1]

### 5.1.2 THE ACQUISITION OF FUNCTIONAL CATEGORIES

The functional categories in examples such as (3) are critical to the correct word order in a language. Languages vary in the order of words in main and subordinate clauses, in a manner that is not illustrated by English. In English, the same order is used in both clauses; in French also, the same basic order applies in both clauses, but in French the tensed verb is raised to a position to the left of an adverbial element:

4 a *John kisses often Mary
b *Jean embrasse souvent Marie*
c John often kisses Mary
d **Jean souvent embrasse Marie*

5

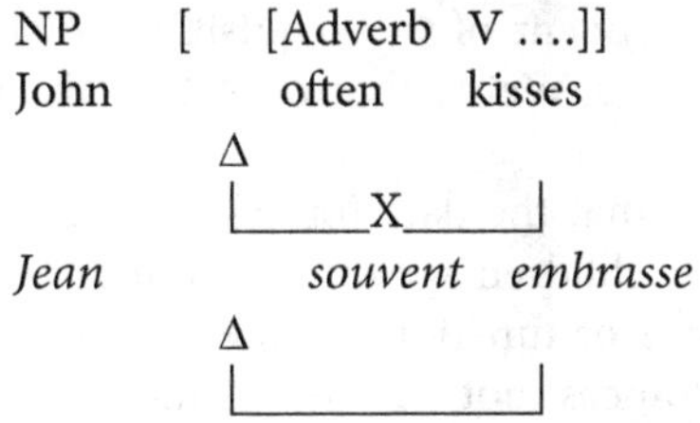

In German, the order of words is different in the main and subordinate clauses. The underlying order is SOV. In main clauses, a verb second rule operates: the tensed verb is placed second, after whatever phrase (subject, adverbial, etc.) occurs in first position. This is illustrated in (6):

6 a *Anton im Garten arbeitet* (underlying order)
Anton in garden works
b *Anton arbeitet im Garten* (surface order)
'Anton is working in the garden'
c *Ich mit meinem Hund spiele, sobald ich nach Hause komme* (underlying order)
I with my dog play as soon I to house come
d *Ich spiele mit meinem Hund, sobald ich nach Hause komme* (surface order)
'I play with my dog as soon as I arrive home'

The movement operation in French and German is held to be triggered by the existence of functional categories. In each case, the movement fills a functional projection above the VP. The movements involve finite (tensed) verbs, and such verbs must agree with their subject. English is not an exception: movement is used to raise the subject NP from within the VP, where it starts out, to the AgrP in (3).

How do children acquire the systems of various languages? An idea that has been popular since the 1980s is that children may go through a developmental stage in which they lack functional categories and projections (Guilfoyle and Noonan 1988; Platzack 1990; Radford 1990; Galasso 2001). We saw in the last chapter that very young children learning languages such as English use 'telegraphic speech'; utterances are produced with the main lexical categories but largely without trimmings such as inflection on verbs. And, thus, it was a natural extension of the child's focus on lexical categories to propose that children at the earliest stages lack functional categories.

How can this hypothesis be tested? The obvious testing ground is word order. Is it the case that very young children may not show the word orders illustrated in (4b) and (6b, d), which result from movement upwards into a functional category, because they lack the mental energy to do so? Or do children fail because they lack the functional categories that trigger movement? Papers in Meisel (1992) take positions on either side of the issue concerning whether there is a very early stage in which functional categories are missing from the child's grammar, and on whether the acquisition of tense is independent of the acquisition of subject-verb agreement (see, for example, Verrips and Weissenborn 1992; Clahsen and Penke 1992).

Much of the problem arises from the fact that the data for assessing children's grammar comes from spontaneous speech by children aged two and under. As just suggested, children may fail to give evidence of functional categories because of limitations on their ability to produce utterances, not because of their non-adult grammars. Valian (2006) used an experimental technique to argue that by age two, children are sensitive to the distinction between present and past tense in English, and we can by extension conclude that TP is a real category in their grammar. Valian studied three contrasts between present and past tense: *will* vs. *did* (7a); copula *be* (*is* vs. *was*, 7b); and progressive *be* (*is* vs. *was*, 7c):

7 a Show me the shoe that I *will/did* tie
  b Show me the bear that *is/was* happy
  c Show me the ball that *is/was* rolling

The experimenter acted out scenarios or asked for a response to picture stimuli. For example, in the case of (7a), two shoes with untied laces were presented; the experimenter declared her intention to tie both of the shoes, and tied the lace of one shoe before asking the child to show her what she would do or had done (a command with *will* or *did*). Responses were coded in terms of the percentage of non-past type responses, i.e. responses in which, for example, the child pointed to the shoe that was untied. Figure 5.1 gives the results for two-year-old subjects for stimuli with no adverb (see below). It is clear that non-past type responses are higher in answer to present tense commands than in answer to past tense commands, with greatest success for *will* vs. *did*, and least success for progressive *be*. But all three conditions showed a contrast. The children's mean length of utterance (MLU) had no effect. A proportion of the children received an adverb in addition to the tense clue to a correct response (for examples such as 7a, the adverb was *next* for present tense and *already* for past; for 7b–c, the adverb was *right now* for present tense and *just before* for past). The two-year-olds in Valian's study showed no effect of the additional clue provided by the adverb, but a group

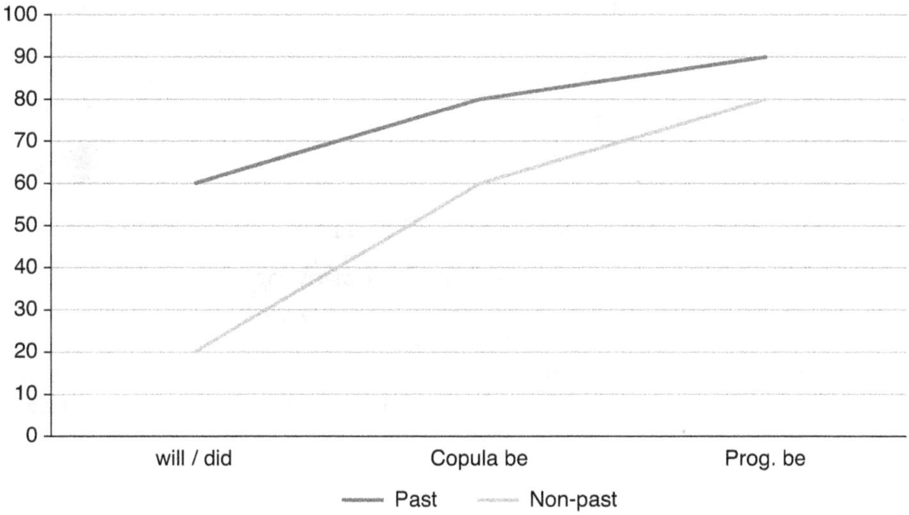


*Figure 5.1 Percentage non-past responses*

*Source:* Valian (2006).

of three-year-olds did benefit from the presence of the adverb in the copula and progressive contexts.

## 5.2 ASPECT AND AKTIONSART

Grammatical aspect refers to the morphological marking and use of auxiliary verbs. We have already seen examples of English progressive aspect. English indicates ongoing activity vs. a state or characteristic property (8a–c) by progressive aspect, or activity that has completed at some point in time prior to the point of speaking (9a–c) by perfective aspect:

8 a Harriet runs (simple present, characteristic activity)
  b Harriet is running (present progressive aspect)
  c Harriet was running (past progressive aspect)
9 a Harriet walked (simple past, characteristic activity)
  b Harriet has walked (present perfect aspect)
  c Harriet had walked (past perfect aspect).

We can draw a distinction between grammatical aspect and *Aktionsart*. Grammatical aspect refers to morphologically and syntactically represented properties of the situation described, such as whether the action is ongoing or completed (as illustrated by examples 8 and 9). Aktionsart refers to properties of the situation that are inherent to the verb involved, particularly whether a change in situation is involved. Verbs such as *break* or *make*, which have an inherent end point, are *telic* verbs; those that do not have such an end point, for example, *laugh* or *sigh*, are *atelic*. Although aspect and aktionsart are distinct concepts, they go hand in hand to the extent that certain verbs, by virtue of their meaning, will lend themselves to use in certain aspects, viz. the progressive aspect is more naturally associated with atelic verbs in English, and the perfective aspect is more naturally associated with telic verbs.

### 5.2.1 CHILDREN CAN DISTINGUISH TENSE FROM ASPECT

Wagner (2001) cites an extensive literature from many languages suggesting that children do not encode adult meanings for tense; rather, it is argued that children may use tense morphology to encode aspect. Wagner terms this the Aspect First Hypothesis, which she tests with an experiment. Her experimental set-up was as follows: a kitten was made to go down a road and perform the same action at each of three places on the road. At the middle place, the child was asked about an event in the past, present or future tense, for example:

10 a Show me where the kitty was hopping around
  b Show me where the kitty is hopping around
  c Show me where the kitty is gonna hop around

Half the verbs were telic and half atelic. Two- and three-year-old children were tested.

The results were as follows: both age groups distinguished each tense from the others, although the difference was more clear-cut for three-year-olds. There was no effect of verb type. This result offers no support for the Aspect First Hypothesis and, taken together with earlier results in Weist et al. (1997) and Weist et al. (1999), and the study by Valian (2006) summarised in the previous section, argues that tense is not conflated with aspect at an early age.

### 5.2.2 THE DEVELOPMENTAL PATH OF ASPECTUAL MEANING

Vinnitskaya and Wexler (2001) found evidence from three- to six-year-old Russian-speaking children that perfective and imperfective (progressive) morphological endings are distinguished. In a study of Dutch, Italian and Polish, van Hout (2008) found that although perfective aspect is acquired earlier than imperfective aspect, there are cross-linguistic differences in the speed of acquisition, with three-year-old speakers of Italian lagging behind their Dutch and Polish peers; she attributes this to the salience of morphological paradigms in the different languages. In addition, a study by Kazanina and Phillips (2007) argue that subtle properties of an adult-like interpretation of aspect may take time to develop. The focus of their study was the 'Imperfective paradox': the fact that telic verbs may be used in progressive aspect. This is illustrated by (11–12):

11 Mary was driving to Washington from Boston
12 While Katy talked on the phone, John was building a toy house
(Examples 2a–b in Kazanina and Phillips)

Predicates such as *drive to Boston* and *build a house* are telic – they have a clear end point, yet they sound perfectly natural with the imperfective in (11–12).

Kazanina and Phillips adapted the technique used by Wagner (2001) to test knowledge of the imperfective paradox in Russian. The experimenter told a story in which a character carried out an activity at each of three landmarks down a road, acting out the events as the story unfolded. At one location only did the character manage to complete the activity. Three- to six-year-old Russian-speaking children and Russian-speaking adults were tested. The experimenter asked a series of questions following the story, using both perfective and imperfective verbs; examples are given in (13) and (14):

13 Gde obezjyanka sobrala gnomika?
Where monkey assemble.Past.Perf smurf
'Where did the monkey build a smurf?'
14 Gde obezjyanka sobirala gnomika?
Where monkey assemble.Past.Imp smurf
'Where was the monkey building a smurf?

Russian adults (including those tested in the experiment) accept only the perfective when the activity was completed – of building a Smurf in the example – but accept both incomplete activities and completed activities when the imperfective was used. In this experiment, Russian children correctly accepted/rejected 95 per cent of the questions with a perfective verb when a completed/incomplete activity was queried; however, their correct responses fell to only 39 per cent when the imperfective was used. The great majority of children's incorrect responses came from a failure to name a location at which the activity was incomplete. Thus, it would seem that children had not mastered the use of the imperfective when a telic predicate was used.

But some children in Kazanina and Phillips's study did show an adult-like pattern of responses, and further experimental work showed greatly improved performance. In a new experiment, *while* clauses were used to set a time interval against which the truth of the main clause was evaluated. Example (15) is judged false and (16) is judged true by adult Russian speakers, following a story in which a boy and a girl are engaged in different activities (watering flowers and cleaning a table), and the girl has not finished cleaning the table at the moment that the boy stopped watering the flowers:

| 15 | Poka | malchik | polival | cvety, | devochka | vyterla | stol |
|---|---|---|---|---|---|---|---|
| | While | boy | water.Past.Imp | flowers | girl | clean.Past.Perf | table |

'While the boy was watering the flowers, the girl cleaned (all of) the table'

| 16 | Poka | malchik | polival | cvety, | devochka | vytirala | stol |
|---|---|---|---|---|---|---|---|
| | While | boy | water.Past.Imp | flowers | girl | clean.Past.Imp | table |

'While the boy was watering the flowers, the girl was cleaning the table'

With the *while* clause determining the time interval, Russian children correctly rejected (15) (91 per cent of responses) and also correctly accepted (16) (89 per cent of responses). These children were also tested on questions such as (13) and (14), and – as in the previous tests with those sentence types – succeeded with (13) and failed with (14). Kazanina and Phillips argue that their findings show that young children do have a grasp of the use of perfective vs. imperfective verbal morphology, but – without a clear temporal boundary set by a *while* clause – they lack the ability to suspend the entailment that an activity is completed when the imperfective is used.

## 5.3 NEGATION

Negation is expressed by a further functional category, which can be analysed as sandwiched between AgrP and TP.[2] A sentence such as (17) will have the structure in (18):

17 Harriet isn't running

18

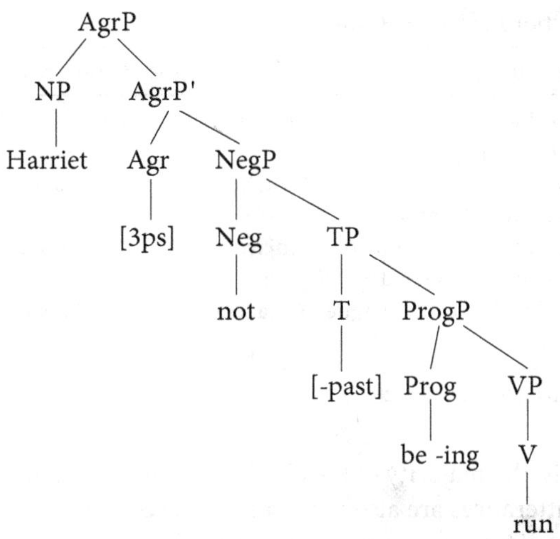


In (18), the verb *be* moves up in the structure to the left of the negative element; this is true also of the auxiliary verb *have* (*Harriet hasn't been running*). If neither the auxiliary *be* nor *have* is present, the dummy verb *do* is inserted (*Harriet doesn't run*).

### 5.3.1 CHILDREN'S NEGATIVE SENTENCES

Drozd (2002) provides a new look at what the child does when s/he creates a negative structure. Many early studies observed that young children produce utterances along the lines of *No X*, where X can be filled by any lexical category in the adult grammar, or by a whole sentence:

19 no good
no flower in there
no crackers for you
no sunny outside
no ready
no over
no Leila have a turn
I no know (Examples from Drozd 2002, (1))

The thrust of analyses ranging from Klima and Bellugi (1973) to Harris and Wexler (1996) has been that the child uses a negative as a general modifier, expressing negativity (or more exactly as a negative operator, having scope over the utterance that follows). Based on a detailed analysis of the negative utterances of ten children, Drozd challenges this view, arguing instead that the majority of child negative utterances are discourse reductions of sentences with *there*, on the lines of (20):

20 No ice cream < there is no ice cream

Table 5.1 Some discourse functions of bare DP ellipsis

| Discourse function | Context and usage | Type of ellipsis | Paraphrase |
|---|---|---|---|
| Existential | A finds no champagne<br>A: *No champagne* | Situational | There's no champagne |
| Imperative prohibition | A warning B<br>A: *No champagne* (for you!) | Situational | There's no champagne |
| Confirmatory denial | A: George drank no champagne<br>B: *No champagne*. Too bad | Adjacency | Previous utterance |
| Recapitulatory question | A: George drank no champagne<br>B: *No champagne?* Why not? | Adjacency | Previous utterance |

*Source:* Abbreviated from the list in table 5, Drodz (2002).

Thus, under Drozd's analysis, the majority of the child's negative utterances are DPs. Other sources of negative utterances are also recognised. For example, in preclausal position (e.g. *No Leila have a turn* or *No mommy cut it*) the negative is treated as analogous to an adult utterance such as *No way mommy cut it* (Drozd 1995).

The attractiveness of Drozd's analysis derives from the establishment of a close parallel to adult utterances, which employ bare DP constructions in a variety of discourse functions, some of which are listed in Table 5.1. Thus, the child's negative utterances can be made to conform to the grammar of the adult, and do not represent a deviation from what is permitted in the adult language.

## 5.4 MODALITY

Modality is a cover term for a range of meanings related to the beliefs and attitudes of the speaker. In English, this range of meanings is expressed primarily by means of modal auxiliaries, which occur as the first of any sequence of auxiliary verbs: *can*, *will*, *shall*, *may*, *must*, etc. Quasi-modal expressions include *wanna* and *gonna*. The meanings of modal verbs are complex and overlapping. A basic distinction is drawn in the literature between *root* and *epistemic* meanings of the modals. (The terminology varies, however: see Lyons 1977; Palmer 1979; Coates 1983.) Broadly, the root meanings convey the speaker's beliefs and attitudes, indicating *inter alia* permission, probability, obligation, intention and ability. The epistemic meanings, related to the root meanings, convey some sense of a process of reasoning – the drawing of a conclusion based on evidence. The two kinds of modal meanings can be expressed by the same auxiliary, resulting in ambiguity. Thus:

21 John must love our leader

is ambiguous between a root meaning ('It is required that John loves our leader') and an epistemic meaning of inference ('Taking the evidence into account, I conclude that John loves our leader').

### 5.4.1 THE ACQUISITION OF MODAL MEANINGS

Stephany (1986), Shatz and Wilcox (1991) and Papafragou (1997) provide summaries of the literature on the development of modal verbs. Stephany reviews data from English and Greek and other languages; Greek uses different means to English for expressing modality, including the subjunctive verb forms (parallel to English *were* in sentences such as *If I were rich, I …*) and modal particles. Although there is quite a lot of variation in the age at which children learning English use modal verbs, there is nonetheless a fairly regular order in which the modals are used, as approximately given in (22) (based on Stephany 1986: 387, figure 1):

22 Wanna/gonna/hafta > can't > won't > can/will > could/would

*May* and *must* are found in only two children of the eight whose data Stephany summarises; the one child who uses *must* is 3;4 on first use.

What meaning does the child have in mind when s/he produces a modal verb? The meanings that emerge first focus on possibility, intention, volition and (in)ability. This is true for both English and Greek. Thus, it appears that the root meanings of modal expressions precede the epistemic meanings. The analysis of infinitives in main clauses in Chapter 3 (the eventivity analysis of Hoekstra and Hyams) draws on such root meanings. Papafragou (1997) suggests that the later use/acquisition of epistemic meanings is related to the cognitive development of the ability to take into account the states of mind of other individuals (the development of a Theory of Mind, see Chapter 6).

There is experimental evidence that even children aged five and older have a modal system that is not wholly equivalent to the adult's. For example, Noveck (2001) studied children's comprehension of the modal *might*. The participants in Noveck's experiment were presented with three boxes, two of which were open and one closed. The open boxes contained in one case a parrot and a bear and in the other case only a parrot. The experimenter told the participant that, 'All I know is that whatever is inside this box [the closed box] looks like this box [pointing to the box with the parrot and the bear] or this box [pointing to the box with only a parrot].' A puppet was then made to say one of the utterances in Table 5.2 (see p. 88). The percentage of correct responses in the table reveals a startling difference between child and adult responses to the puppet's utterance that there *might be a parrot* in the box. Children in all age groups say that the puppet is right in the large majority of their responses, whereas adults say that the puppet is wrong in the large majority of responses. The difference between the child and adult responses is neatly summed up in the title of Noveck's article: 'When children are more logical than adults'. There clearly is a parrot in the box, because both of the open boxes contain a parrot. Adults use a *scalar implicature* to judge the statement: *might be x* is weaker than *must be x* and adults reject the statement that there *might be a parrot* because they prefer to use *must be a parrot*. Children, on the other hand, accept the statement *might be a parrot* because (we assume) they do not compute this implicit contrast.

Table 5.2 Percentage of correct responses to what was in the closed box

| Presented statement | Correct response: Is the puppet right? | Age (number of subjects) 5 (32) | 7 (20) | 9 (16) | Adult (20) |
|---|---|---|---|---|---|
| Has to be a parrot | Yes | 75 | 90 | 88 | 100 |
| Does not have to be a parrot | No | 72 | 75 | 75 | 100 |
| Might be a parrot | Yes | 72 | 88 | 69 | 35 |
| Cannot be a parrot | No | 66 | 80 | 100 | 83 |

*Source:* Adapted from Noveck (2001: table 2).

## 5.5 CHILDREN'S KNOWLEDGE OF QUANTIFICATION

The set of quantifiers in English includes *all*, *each* and *every*. The comprehension of these words has been the subject of many studies. We will begin with one that demonstrates sophisticated knowledge of the universal quantifier *all*.

### 5.5.1 THE RESULTS OF BROOKS AND BRAINE

Two experiments by Brooks and Braine (1996) showed that children did confine their interpretation of quantifiers to the noun phrase that contained them (contrary to previous analyses, see below), and that they were sensitive to the syntactic context in which the quantified NPs were embedded. Both experiments used a task in which the children had to choose which of two pictures fitted a stimulus sentence. In their first experiment, Brooks and Braine presented children with sentences such as those in (23a–b):

23 a All of the men are carrying a box
   b There is a man carrying all of the boxes

*Figure 5.2 Images to elicit the reading of* all *(experiment 1, Brooks and Braine 1996)*
*Source:* Brooks and Braine (1996).

Table 5.3 Percentage correct responses

| | Age | | | | | | | |
|---|---|---|---|---|---|---|---|---|
| Sentence type | 4 | 5 | 6 | 7 | 8 | 9 | 10 | Adult |
| All of the Xs are (verb)ing a Y | 83 | 97 | 100 | 100 | 100 | 100 | 100 | 100 |
| There is an X (verb)ing all of the Ys | 83 | 90 | 97 | 100 | 100 | 100 | 100 | 100 |

*Note*: The percentages in this table and in Table 5.4 have been rounded to the nearest whole number.

For (23a–b), the child had to choose between the pictures in Figure 5.2. The correct answer for (23a) was the left picture in Figure 5.2, and the correct answer for (23b) was the picture on the right. There were ten subjects in each age group. The results are given in Table 5.3. For all child groups, performance was very good, or perfect, for *all.*[3]

In a second experiment, children were tested on sentences in which the syntax was either active or passive, as given in (24a–d):

24 a All the men are building a boat
b A boat is being built by all the men
c Each man is building a boat
d A boat is being built by each man

Examples of the pictures to choose from are given in Figure 5.3 (p. 90). The top picture shows a collective activity – the men are all working on a single boat; the picture below shows a distributive interpretation – each man is working on a separate boat.

In this experiment, there was no correct answer. Linguistic research (Ioup 1975) has shown that there are different preferences according to the syntax of the clause. If the clause is active, there is a preference for the distributive interpretation; if the clause is passive, a collective interpretation is preferred. The results in Table 5.4 bear this out. Fifteen subjects in each of the child groups aged four, seven, eight and nine, and twenty subjects in the remaining groups were tested.

The results of Brooks and Braine are important for two reasons. First, experiment 1 demonstrates that the quantifier *all* is restricted to the noun phrase it modifies. If that was not the case, then it would not be possible to choose which picture was correct, since that choice involves ignoring the extra boxes in the picture. Second, experiment 2 bears on whether children create more abstract structures than the surface string of words suggest. Movement is not restricted to overt movement. In addition to the formation of questions and passives and other sentence types in which the underlying order is disrupted, syntactic theory posits covert movement. Such movement applies to create structures that are not represented in the surface syntax. The ambiguity of sentence (24a) can be represented in terms of an operation – Quantifier Raising – that produces two structures (approximately (25) and (26)), one of which has the quantifier *all* in a higher position than the existential operator (corresponding to *a*), and the other with the quantifier *all* in a lower position:

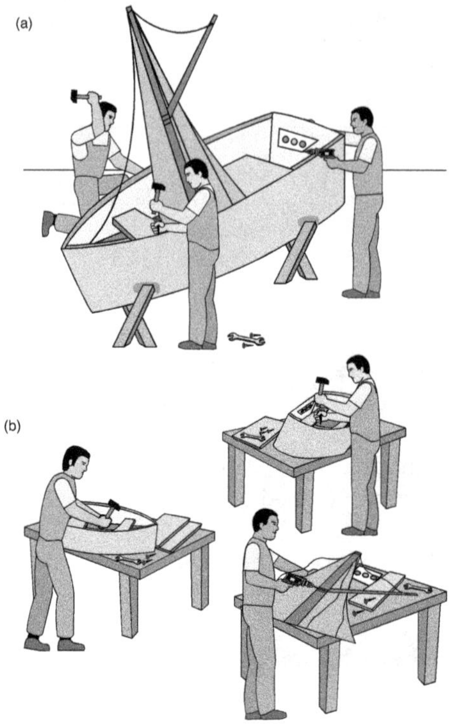


*Figure 5.3 Images to elicit collective vs. distributive readings (experiment 2, Brooks and Braine 1996)*

*Source:* Brooks and Braine (1996).

Table 5.4 Percentage collective interpretations

| | Quantifier | | | |
|---|---|---|---|---|
| | *All* | | *Each* | |
| Voice: | Active | Passive | Active | Passive |
| 4-year-olds | 54 | 57 | 38 | 62 |
| 5-year-olds | 57 | 83 | 44 | 72 |
| 6-year-olds | 69 | 93 | 27 | 52 |
| 7-year-olds | 76 | 98 | 24 | 47 |
| 8-year-olds | 71 | 94 | 1 | 38 |
| 9-year-olds | 72 | 97 | 7 | 23 |
| Adults | 83 | 98 | 1 | 18 |

25

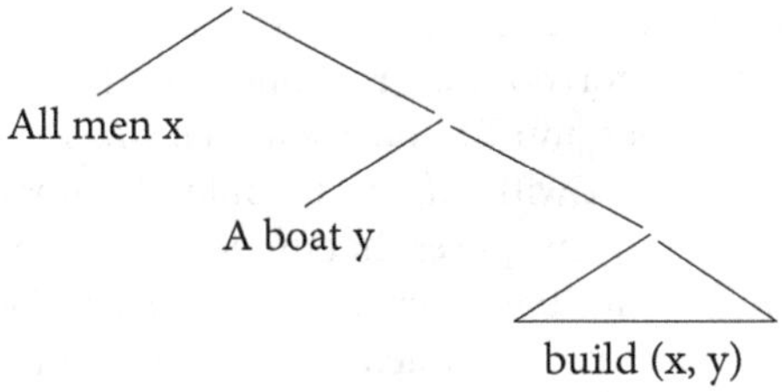


26

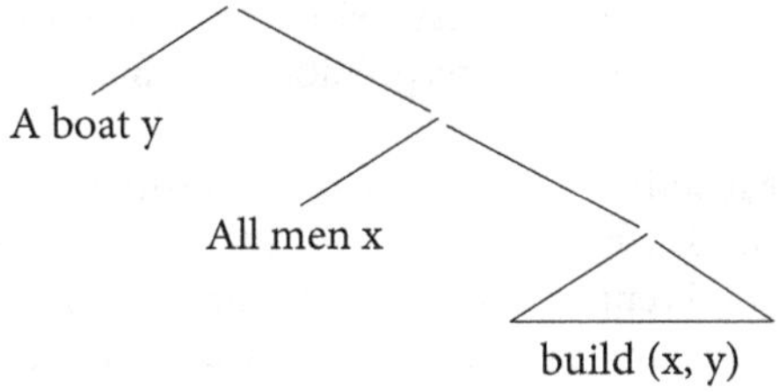


Example (25) corresponds to the distributive interpretation and (26) to the collective interpretation. Not only do the results for experiment 2 suggest that children do build structures that result from Quantifier Raising; the effect of passive syntax suggests that children, like adults, prefer interpretations in which the structures that result from Quantifier Raising correspond to the surface order of constituents in the input sentence.

### 5.5.2 YOUNGER CHILDREN

Although from age five to six onwards the adult pattern of interpretation is found in Table 5.4, the four-year-olds do not reflect performance that is similar to adults. What kind of representation(s) do children under five have for quantifiers?

Experiments with children younger than five have revealed patterns of error that may reflect not only the quantifier involved, but the mapping between the experimental set-up and the response. Drozd (2001) summarises differences between adult and child responses found in various studies (Inhelder and Piaget 1964; Donaldson 1978; Philip 1996). Faced with pictures of various combinations of boys and elephants, adults will reply 'Yes' to the situation in answer to the question *Is every boy riding an elephant*?, in which the picture shows all the boys on a different elephant, with an elephant left over. They will also answer 'Yes' to a situation in which all the boys are on the same elephant, again with one or two 'spare' elephants. Frequently children aged three to four will say 'No' in both cases, citing the existence of the spare elephant(s) (but children also give correct answers). By contrast, adults will say 'No' to a situation in which a number of boys are riding an elephant, with a boy left over (not riding an elephant), whereas young children will say 'Yes', ignoring the spare boy.

There are several accounts – not necessarily in competition – of the errors young children make: Philip (1996) proposes an 'event quantification' account, by which the quantifier is interpreted as modifying the entire predicate; that is, the child interprets

*Is every boy riding an elephant?* as if it were the question *Is it always the case that every boy is riding an elephant*, which will account for both correct and incorrect child answers. (The results of experiment 1 by Brooks and Braine argue that this is not a good explanation for children aged four and over with *all*, but it could be a possible analysis for younger children.) And Drozd (2001) proposes that universal quantifiers such as *all* and *every* are interpreted as weak quantifiers such as *many* or *a few*. Weak quantifiers permit a range of numerical values: for example, *many* can mean ten or 100, depending on the comparison group. Thus, the weak quantifier analysis, like event quantification analysis, can account for both exhaustive interpretations (all the boys are engaged in an activity) and non-exhaustive interpretations (some boys are not engaged in the activity).[4]

Brooks et al. (2001) and Lidz and Musolino (2002) contribute cross-linguistic perspectives to the debate. Brooks et al., drawing on the work of Vendler (1967) and Ioup (1975), propose that in the mental representations of children both collective and distributive meanings are present, but that the two are variably accessible, due to the realisation of quantification in the language the child is learning. For example, in both Mandarin Chinese and Portuguese, unlike in English, the indefinite article is homophonous with the numeral *one*, and in both these languages, collective interpretations for children are more easily accessed than for English children of the same age.

Lidz and Musolino studied the interpretation of negative sentences with NPs modified by a numeral (one type of quantifier) in English and Kannada, a Dravidian language spoken in south-western India. Kannada has a different basic word order to English, with the object preceding the verb. In both these languages, there is an ambiguity in sentences such as the English (27) and the Kannada (28):

27 Alex didn't wash two cars
28 Anoop eradu kaaru toley-al-illa
two car wash-inf-neg

The sentences can be interpreted as meaning either that there are two cars that person A did not wash, or that it is not the case that person A washed two cars. This can be captured by representation similar to those in (25–26), with the quantifier *two* in the top position and the negation in the lower position (for the first reading), and the opposite arrangement for the alternative reading. In a Truth Value Judgement Task, in which participants were asked to judge the veracity of stories acted out with props, both English-speaking and Kannada-speaking adults could access the two readings of sentences such as (27–28). By contrast, four-year-old children in both languages strongly preferred the reading under which there were only two cars that were not washed, and dispreferred the reading under which it is denied that only two cars were washed. One interpretation of these results is that children rely on local relations in their comprehension of sentences: the negation is adjacent to the verb in both languages, and the preferred interpretation by children is the one in which the verb is negated (see Exercise 3 for further thoughts on this).

## 5.6 MORE CONNECTIONS BETWEEN LEXICAL STRUCTURES AND MENTAL STRUCTURES

The availability of mental representations and lexical items to represent them has been studied in other areas. Here we will focus on two: over- and under-extensions of meaning, and the distinction between count nouns such as *books* (plural *books*) and mass nouns such as *water* (ungrammatical plural **waters*, except in special uses).

### 5.6.1 OVER- AND UNDER-EXTENSIONS

Clark (1973) examined many cases in which children, typically aged between one and two-and-a-half years, over-extended the meaning of words, in a way that can be described in terms of a focus on perceptual features common to sets of lexical items. For example, the word *moon* may be used to refer not only to the moon, but to all kinds of round objects. Griffiths (1986) argued that under-extensions of a word's meaning, although more difficult to spot, may be quite a normal and frequent stage in the development of the meaning of words. For example, the word *dog* may be used only to refer to the family's pet, not to canines in general. Griffiths argues that a word is associated with a mental image, derived in more or less detail from the perceptual event – the experience – from which the child has learned the word. The child must learn from experience that a more general reference is appropriate.

More recent research has confirmed that both over-extensions and under-extensions are not uncommon in the development of word meanings, but that misuses interact with the structure of the language being acquired, limiting their role in development. Bowerman and Choi (2001) focus on the acquisition of spatial terms. Bowerman and Choi (summarising previous work on several languages) demonstrate that from the earliest ages (under eighteen months), children are attuned to the language-particular ways of expressing spatial relations. English uses prepositions such as *on* (for example, *put a book on a table*) and *in* (*put a piece in a jigsaw puzzle*). A basic distinction in English is thus that of placing an object into contact with another object vs. putting an object into some kind of enclosure. Korean uses different verbs, reflecting, for example, whether or not containment is loose (a pillow in a case) or tight (a piece in a jigsaw puzzle). English-speaking infants and Korean-speaking infants immediately latch on to the method of expressing spatial terms in their language (prepositions vs. verbs). When they made mistakes, there was a common tendency to cluster the mistakes around cases where the boundary between uses in the particular language is not completely clear. For example, a child learning English may use *in* for placing a ping-pong ball between two knees (which is perceptually an enclosure of sorts), and a Korean-speaking child may use the verb *kkita*, meaning 'fit tightly', for sticking a fork into an apple (for adult Koreans another verb would be used, because the holes did not exist before the action took place).

### 5.6.2 COUNT VS. MASS NOUNS

English has a system that syntactically distinguishes two classes of nouns: count and mass. Count nouns such as *book* can occur in both singular and plural forms, and can be modified by numerals, and by determiners such as *several* and *many*; mass nouns such as *water* do not have plural forms, cannot be modified by numerals, and can be modified by terms such as *little* or *much*. These properties are illustrated in (29):

29 a a/one book
b *a/one water
c two books
d *two waters
e many books
f *many waters
g *much book(s)
h much water

Count and mass nouns on the face of it seem to be distinguished in terms of whether they pick out definable, perceptual individuals or whether they correspond to 'blobby' things, substances with no clear individuation. However, there are exceptions to this generalisation. For example, the words *fruit* and *silverware* denote perceptually discrete classes of objects but are mass nouns in English (**two fruits*, **two silverwares*). Philosophers and psychologists have puzzled over whether there is a stage in which young children can make perceptual distinctions prior to linguistic distinctions or whether the acquisition of, for example, pluralisation of nouns leads to intellectual growth in the area of perception of things as objects. Quine (1960) is an example of the latter type of thinking and Macnamara (1982) of the former.

Gordon (1985a) tackled the question of the existence of a stage in which objects are classified in terms of their perceptual properties rather than their grammatical properties (count vs. mass). Gordon reasoned that if children have word categories that are not syntactic, but are rather based on semantic and perceptual classes, then we would expect them to miscategorise exceptions to the general case that count nouns denote discrete objects and mass nouns denote substances. If they do not make such errors, then it is plausible to conclude that words are classified in their grammars in a way that cannot be reduced to semantic, non-syntactic properties. In one of Gordon's experiments, children aged two to five responded to questions about shops and shopping – for example:

30 Do you know what you get in the {fruit/vegetable} section?

The adult answer is a plural form for the count noun (*vegetables*, not **vegetable*) and a singular form for a mass noun (*fruit*, not **fruits*), despite the fact that the mass noun has perceptual properties associated with count nouns. For children who knew the words involved, there were very few errors in which children treated the mass

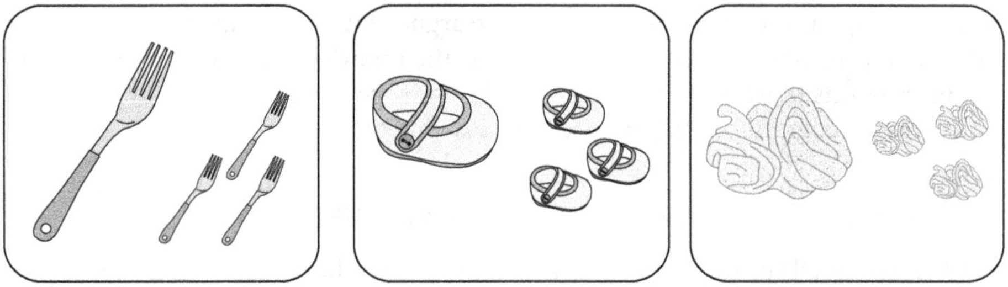

*Figure 5.4 Images to test mass and count syntax*

*Source:* Barner and Snedeker (2005).

nouns as count, incorrectly pluralising, and there were no such errors of this type for two-year-olds.

Barner and Snedeker (2005) deepen our understanding of the results by Gordon. In one experiment, they compared performance of adults and four-year-old children with three classes of nouns: count nouns such as *shoe*, object-mass nouns such as *silverware* and substance-mass nouns such as *toothpaste*. Object-mass nouns are mass nouns that refer to perceptually discrete entities, and substance-mass nouns refer to entities that do not. The participant was asked *Who has more X?* when confronted with two characters, one of whom had one large entity and the other had three small entities (see Figure 5.4). The results were clear-cut. Both adults and children responded that the character with more for both the count nouns and the object-mass nouns was the one with three entities (100 per cent and 97 per cent for adults, and 89 per cent and 95 per cent for children), and the character with more for substance-mass nouns was the character with one large entity (100 per cent for adults and 91 per cent for children). In another experiment, Barner and Snedeker looked at nouns that can be used as either count or mass, such as *string* and *stone*. The nouns were presented in either count or mass syntax (*Who has more stones?* or *Who has more stone?*) and the same options were offered, one large entity or three small ones. Both the adults and children used syntax to guide their judgements, choosing the character with the greater number of entities when confronted with count syntax (plural) and the character with a single object of greater size than the three objects combined when confronted with mass syntax (absence of pluralisation).

## CHAPTER SUMMARY

This chapter has covered a number of topics in the development of meaning. Recurring threads have been whether children exhibit the same range of meanings that adults exhibit and how children map between syntactic structure and meaning. Sometimes the meanings given by children are more restrictive than those given by adults (for example, in children's interpretation of modals) and sometimes the interpretations are different/more liberal (for example, in younger children's interpretation of quantifiers). Nonetheless, we can argue that the system cannot be wholly reduced to properties of general cognition, as the knowledge of the syntax of count and mass nouns illustrates. The relative independence of the linguistic system from other areas of cognition is taken up in the next chapter.

## QUESTIONS AND EXERCISES

1. We mentioned (p. 80) that subjects are raised from the VP into a functional projection. What – if any – prediction does that make for the hypothesis that lexical projections emerge in a child's grammar earlier than functional projections? Suppose the base structure for a VP in a language such as English is [$_{VP}$ NP$_{SUBJECT}$ V ... ]. What kind of data might inform this question?

2. In Chapter 5 we saw that a child produced the utterance *Mummy must have gone shopping* at 24 months. It appears that she was using the modal *must* epistemically. What can be said of this, in the light of observation that root uses emerge before epistemic uses of modals? For example, was the child an exceptionally precocious learner? Does a child's environment work against epistemic uses of modals?

3. Lidz and Musolino propose an account in terms of c-command, based on the following structures:

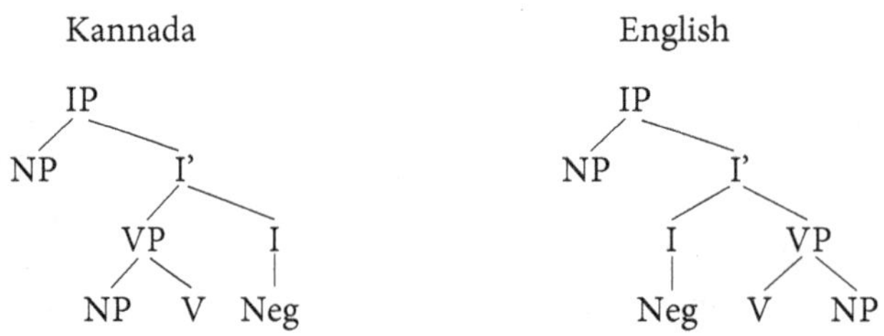


In both of these structures, Neg c-commands the VP. Notice that the structure assumed for Kannada is more abstract than the actual utterances in Kannada, in which the negation is attached to the verb as a morpheme.

Ask yourself the following question: is the processing account given in the text (based on adjacency of the negation to the verb) of the fact that children prefer the interpretation 'There are two cars that A did not wash' better than the c-command account that Lidz and Musolino propose, or do they boil down to the same thing?

## NOTES

1. The precise system of English morphological agreement is not a settled matter; however, in all analyses movement of features (for person/tense) is involved.
2. The ordering of functional phrases above the VP is a matter that analyses differ on. Haegeman and Guéron (1999) place the NegP above TP, whereas Adger (2003) places it below. We follow the analysis adopted below by Drozd (2002), but this is not important to

Drozd's main points. Roberts (2018) provides a typology of variation in the positioning of Neg.

3. Experiment 1 in Brooks and Braine's study also tested *each*, with performance that was considerably worse for children aged six and under.
4. Crain et al. (1996) suggest that the errors children make may be enhanced by the experimental procedure/set-up; however, this does not rule out an incorrect analysis when the child does make errors.

# CHAPTER 6

# COGNITION, ENVIRONMENT AND LANGUAGE LEARNING

In this chapter we will take up contentious questions concerning how the child develops knowledge of grammar. Over the past twenty years old problems have been revisited, and claims concerning the nature of children's knowledge and how the child acquires it have been subjected to fresh scrutiny. This chapter makes the argument that challenges to the claim that language acquisition is guided by an innate knowledge of Universal Grammar are inadequate to account for the facts of development. First, we will review traditional arguments for innate knowledge of grammar. Second, we will consider whether or not the alternatives proposed really are alternatives. Then, we will examine the question of how language develops in special populations and how grammar development may drive cognitive development or vice versa. Finally, we will take a look at the some of the most recent work in learnability theory – how the child can use her/his input to acquire the fine points of the grammar s/he is exposed to.

## 6.1 INNATENESS

What does it mean to say that the child has innate knowledge of Universal Grammar? Clearly, it can't be that the rules of any particular language are biologically given. Rather, it has to be the case that the general patterns which set limits on what a language can be like are part of the child's predisposition to learn. Arguments for innateness fall into two types. The first type draws a parallel with other behaviours characterised by biological programming. The second type points to the relative independence of what is acquired from external stimuli.[1] Both types of argument can be made for the development of language.

### 6.1.1 AN ORDERLY PROGRESSION OF STAGES

In different areas of grammar, we have seen that there is a distinct sequence for the development of linguistic abilities. For example, babbling emerges before first words; a one-word stage generally precedes the emergence of multi-word utterances; and specific morphemes emerge in approximately the same order across children learning the same language (as, for example, in the order of emergence of English morphemes in Brown's 1973 study). Although the age at which a particular stage is reached may vary considerably, there are rough guidelines (for example, the babbling period is generally placed between six months and the turn of the first year). A pattern of ordered stages in development is one of the hallmarks of biologically programmed behaviours (Lenneberg 1967). The existence of regular stages in language development is analogous to other biologically triggered phenomena in both humans and other species (such as walking in humans and flight in birds).

### 6.1.2 A CRITICAL PERIOD

When children are deprived of normal linguistic stimulation until they are around age six, they fail to fully develop language. The most famous cases are summarised in Gleitman and Newport (1995). Isabelle was isolated and never spoken to by a mentally-ill mother. She was discovered at age six, having learned no language and with the cognitive development of a typical two-year-old. Within one year of her discovery, she had developed language at the level typical of her age group (Davis 1947). By contrast, Genie was discovered at thirteen years of age, having also been isolated and subjected to appalling cruelty. Despite intensive therapy, Genie's language never approached the normality of an adult speaker. She acquired a substantial vocabulary and could compose meaningful sentences (for example, *Another house have dog* and *No more take wax*, Gleitman and Newport 1995), but she never developed the complex syntactic abilities that we have seen to characterise the ability of a typical three-year-old (Curtiss 1977). A critical period is one of the hallmarks of genetic programming – for example, if sight is impaired during early life either for medical reasons or (in non-human animals) by experimental manipulation, the ability to see will permanently be damaged.

### 6.1.3 INDEPENDENCE FROM INPUT

Development is to a degree independent of external stimuli. Clear evidence of this is the fact that deaf children engage in oral babbling (Locke 1983: chapter 1). Such activity must be the result of a biologically timed programme that is not dependent on exposure to speech.

Even more striking is the spontaneous development of language by communities in which the input is highly limited. The Nicaraguan and Al-Sayyid Bedouin sign languages developed when people who were deaf had the opportunity to interact with other deaf people; a sign language developed which displays rule-governed properties distinct from the spoken languages around them (see Kegl et al. 1999 for Nicaraguan sign language and Aronoff et al. 2008 for Al-Sayyid Bedouin sign language).

Creole languages evolve from pidgins, which are highly restricted systems used for communication between mutually unintelligible languages, usually for purposes of trade. Creoles exhibit certain characteristic properties – such as particular tense and aspect systems. Although there is a great degree of controversy concerning the role of child and adult learners of creoles (see DeGraff 1999), it is nonetheless the case that such languages display properties that do not derive from the parent (input) languages. Such properties must derive from the minds of the speakers who create them.

The existence of a biological programme for language learning does not preclude a vital role for external stimuli. As a case of extreme deprivation such as Genie shows, a child has to be exposed to language for normal development to take place. It is generally agreed that the child is an active learner, who 'works on' the speech s/he hears, using some kind of grammar-forming mechanism. Current views of the nature of that mechanism have been developed in the context of several facts and assumptions about the nature of the learning situation. These are the topic of the next sections.

## 6.2 INPUT AND ERRORS

*Input* is usually understood to mean the speech forms that the child is exposed to, which may be augmented by contextual clues as to what an utterance means. The nature of the input itself provides a strong argument for the position that the child is equipped with a highly structured grammar forming mechanism.

Input can be divided into *positive* and *negative* evidence. The positive evidence (or positive input) is evidence that a particular form exists in a language. One important limitation on positive input is that the sentences children hear do not contain overt information about their structure and meaning. Sentences do not come ready tagged with information about what the component parts of the sentence are, or what the sentence can or cannot mean. Some refinement of this point is in order. Research over the past twenty years has argued that infants are sensitive to statistical regularities in the speech stream (Saffran et al. 1996; Saffran and Thiesson 2003; Lew-Williams and Saffran 2011). For example, Saffran et al. (1996) found that infants aged eight months were sensitive to the transitional probabilities (the probability of one sound following another) in nonsense words; they listened longer to strings that were unfamiliar to them from a training phase than those words that were familiar. Since transitional probabilities are lower between words than within words, this is an important cue for identifying word boundaries. Such results complement the sensitivity to prosodic boundaries in infants (p. 74 above). Despite such findings, the basic point remains that when the child has identified a word in her/his language, no information is present concerning its identity and structure, other than whatever situational cues are present (on the limits of such cues, see section 6.8.2 below). Another limitation on positive input is that the child hears only a limited sample of the sentence types that are actually grammatical in her/his language.

Not only do children generally get no overt information about structures and meanings, they are also not informed about which strings are ungrammatical. There is a lack of negative evidence. Brown and Hanlon (1970) found that parents responded to the truth value of children's utterances but did not overtly correct ungrammatical forms produced by a child (see also Demetras et al. 1986). Marcus (1993) robustly argues in favour of Brown and Hanlon's position, and against claims that, for example, parental responses to child utterances contain clues to grammaticality.

The restricted nature of the input (the paucity of positive evidence and the lack of negative evidence) is referred to as the *poverty of the stimulus* and has frequently been invoked by Chomsky and others as a point in favour of a learning mechanism in which the child's innate knowledge of principles of grammar plays a major role in guiding development (see Chomsky 1959, 1965; Hornstein and Lightfoot 1981).

Poverty of the stimulus arguments have been challenged. For example, Pullum and Scholz (2002) have questioned the lack of evidence for the structure dependence of linguistic rules: one has to invert the main clause auxiliary to form a question, not the first auxiliary – (1b) is grammatical, not (1c), as the question equivalent to (1a):

1 a The dog that is in the corner is hungry
  b Is the dog that is in the corner hungry?
  c *Is the dog that in the corner is hungry?

Pullum and Scholtz point to the fact that data supporting the inversion of the main auxiliary as opposed to a subordinate auxiliary is present in a non-trivial amount in corpora they examined. However, their argument is flawed for two reasons. First, Legate and Yang (2002) observe that the data Pullum and Scholtz cite come from written (newspaper) texts; when Legate and Yang examined data from child-directed speech, the proportion of relevant utterances was much lower (less than .07%). Second, Lasnik and Uriagereka (2002) observe that even if the data that supports a structure dependent rule (invert the main auxiliary) is present in a (very small) proportion of utterances, this does nothing to rule out any number of competing incorrect hypotheses. This point is reinforced by the fact that we have very robust intuitions about the grammaticality and ungrammaticality of sentence types such as (2a) vs. (2b), for which adults have no evidence or vanishingly little evidence in the input:

2 a These are the reports that I filed _ before reading _?
  b *These are the reports that I filed the papers before reading _?

In (2a) there is a gap formed by movement after *filed*, and this licenses a second gap (a 'parasitic' gap) after *reading*, from which position movement is not ordinarily permitted (2b); see Cowart (1997) for experimental evidence from adult subjects confirming the distinction between sentences such as (2a) vs. (2b). Gil et al. (2018, pp. 55–64) summarise experiments using child subjects in favour of the poverty of the stimulus.

## 6.3 THE ROLE OF UNIVERSAL GRAMMAR IN LANGUAGE DEVELOPMENT

We will assume that the child analyses input sentences into a string of words via a sentence processing mechanism. (The exact nature of the processing mechanism – both for children and adults – is a complex and vexed issue and is the topic of the next chapter.) Such strings will then be analysed by existing rules of the language being learned and/or principles of Universal Grammar. The primary role of Universal Grammar in language development is to limit the hypotheses that a child can form concerning the rules of her/his language, thus also limiting errors and helping explain the speed and ease with which language is learned. Moreover, Universal Grammar has the potential to set off a chain of hypotheses that take the child beyond the sentences in the input. For example, consider a simple question:

3 Will Mary choose something unusual?

The inversion of the subject and the auxiliary *will* may cue the child to the use of movement as an operation. Once set in motion, the use of a movement mechanism may spread to other constructions, including the placement of *what* in the example (4):

4 What will Mary choose?

The connection between subject-auxiliary movement and wh-word movement is not entailed by the theory of movement itself. Rather, this is an example of how simple data such as the string *Will Mary choose …* can in principle trigger far-reaching consequences in the child's grammar; movement may 'spread' from one construction to another without overt prompting from the speech in the child's environment (cf. Lightfoot 1989, 2017).

## 6.4 LEARNABILITY AND ACQUISITION PRINCIPLES

*Learnability theory* is concerned with the conditions under which successful learning of a system can take place within a finite amount of time. The terms *learnability* and *learnability theory* were originally associated with work on learning formal languages, which may conform only to a degree with natural language systems (Gold 1967; Wexler and Culicover 1980; Wexler 1981). But they have come to be used in a more general sense, to refer to the study of conditions that will permit successful learning of natural languages within a limited time span. In this more general sense, learnability theory takes in questions such as the relative contribution of principles of Universal Grammar and input (the speech forms the child hears) in grammar formation, the prevention of errors and the correction of errors. For a recent review of issues in learnability theory, see Fodor and Sakas (2017).

One aspect of learnability theory has been a putative principle governing the child's hypothesis formation – the *subset principle* (Berwick 1985; Wexler and Manzini 1987). This was a principle designed to prevent the learner falling into error when more than one analysis is permitted under the principles of grammar. The child's grammar is hypothesised to permit only those rules for which s/he has positive evidence. For example, although in English a reflexive pronoun must refer to an NP inside the same clause (*Fred* in 5),

5 Tom said that Fred had shaved himself

other languages have more complex systems, and permit reference outside the clause under certain circumstances, as we have seen in Chapter 4. Korean is one language which permits reflexives to refer to a higher subject. The subset principle predicts that the child will choose to interpret a reflexive as co-referential with the subject of the lower clause, until the point where s/he receives positive evidence that co-reference outside the clause is permitted. Lee and Wexler (1987) reported that children learning Korean preferred the structurally closest antecedent for a reflexive (contrary to adults, who preferred a more distant antecedent). However, the empirical evidence in favour of the subset principle, with its intention to prevent error, is not strong. Cho (2009) reports evidence from Korean that goes against the predictions of the subset principle, and points to lexical (verb-based) restrictions on the acceptability of reflexive co-reference. And similarly, we have seen the case of Danish, which requires a reflexive to refer to a higher subject under lexically specified conditions (see examples (41a–b) in Chapter 4). Jakubowicz (1994) found that error rates for the reflexive form that requires long-distance binding were high until age seven and older. We return to the question of lexically restricted phenomena in section 6.9 below.

Although the subset principle has not fared well in the face of child language data, another candidate principle appears to be operative in adult languages, and potentially restricts children's hypothesis formation: the principle of unique function.[2] This principle roughly states that other things being equal, the grammar favours non-ambiguity of meaning or reference. For example, the words *gaiety* and *gayness* have distinct meanings in English: *gaiety* meaning the mood of light-heartedness and *gayness* homosexuality. *Gayness* was introduced into the language because *gaiety* already had an assigned meaning. Clearly this is a tendency, not an absolute rule – otherwise we would have no ambiguity in language. But the avoidance of ambiguity may drive a child's hypothesis formation. Consider one example given in Chapter 3. Clark's child distinguished two different meanings for adjectives in English: he used the ending *-y* to indicate inherent states, and the ending *-ed* to indicate temporary states.

## 6.5 SUMMARY: COMPONENTS OF A LEARNING MODEL

We can summarise the points made in the previous sections with a diagram (Figure 6.1). The input to the learning mechanism is analysed by the sentence processor.

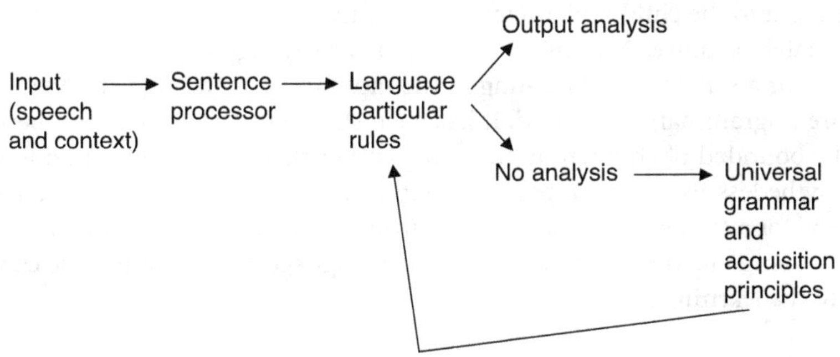


*Figure 6.1 Components of a learning model*

If the learner's current language-particular grammar contains no rules suitable for the input, the input string is subjected to analysis by Universal Grammar and acquisition principles.

## 6.6 SOME PUTATIVE CHALLENGES TO THE MODEL

Before looking in more detail at components of the learning model, including the model's shortcomings, let's delve into other putative challenges to the claim that the child comes to language learning with a knowledge of Universal Grammar, additional to the criticisms of the poverty of the input sketched in section 6.2.

### 6.6.1 MOTHERESE

In some of Chomsky's early writings, the existence of ungrammatical, fragmentary input was mentioned as enhancing the plausibility of a strong innate component of linguistic knowledge. Adults do not always talk in perfectly well-formed sentences, yet any slip-ups adults make do not seem to hinder language acquisition. In 1972, Catherine Snow published a study that challenged the factual basis for the claim that speech to children was rich in potentially misleading errors and disfluencies. Snow found that mothers' speech (more accurately, adults' speech) to children has special features that set it apart from their speech to other adults or to older children. Mothers' speech forms are characteristically fluent and intelligible, with fewer hesitancies and fewer mumbled or muffled utterances than speech to adults; moreover, there were very few utterances than were not well-formed, and mothers' utterances were shorter to children (the average length of utterances to two-year-olds in Snow's study was about 6.5 words, as compared to 9.5 in the speech to ten-year-olds). This special style of speaking to children was given the name *motherese*. The existence of motherese was widely taken to diminish the force of the innatist position, with its inbuilt knowledge of Universal Grammar (see Brown 1977: 20).

There were (and are) two basic responses to the supposed efficacy of child-directed speech. First, motherese in no way diminishes the force of the 'poverty of the stimulus' argument for innate knowledge of UG (section 6.2 above). The fact that adults use short, clear, grammatical forms does not in any obvious way answer the point that the input to the child underdetermines the nature of the grammar that the child will ultimately acquire. Nothing in the input overtly signals the range of possible meanings for a sentence and nothing in the input directly tells the child which sentences are ungrammatical. Second, it has been shown that the use of motherese is a culturally bounded phenomenon. In settings other than the Western middle classes there may be less use of the types of speech forms characteristic of motherese (see Snow 1986 for a review). Since all children from all classes and cultures normally end up with a complete knowledge of their native language, use of motherese cannot be critical to the learning process.[3]

### 6.6.2 USAGE-BASED APPROACHES TO ACQUISITION

The research into motherese as a viable alternative to innate knowledge of principles of Universal Grammar pretty much died out in the 1980s (although research into infant-directed speech continues with other goals). But fascination with the idea that a child can learn language without the benefit of implicit knowledge of UG did not. Since the 1990s, the literature that purports to challenge an innate UG has burgeoned in various manifestations. The terms *emergentist* and *constructivist* are commonly used to refer to this literature.[4] The notion is that grammar develops as an outgrowth of language use. See Tomasello (2003) for a prominent example of usage-based theory. We consider here some general objections, profoundly problematic to usage-based theories.

One basic problem is that raised in the previous section: usage-based theories do not provide an adequate response to the argument from the poverty of the stimulus. The child must acquire knowledge of a system that is not represented directly in the input data. Moreover, the child is creative. We have seen cases in the previous chapters of the child innovating a system that is not represented in the language at hand. Again, the example of Clark's child is relevant (section 6.4 above). And I found positive experimental evidence in favour of the hypothesis that preschool children use a nominal analysis for controlled adjunct clauses (p. 70 above). Such facts are readily accounted for under a theory that gives the child access to the range of permissible systems in languages of the world. They are not in any obvious way explicable in terms of the child working on the basis of the input.

Fodor (1980) examined the logic of learning, arguing that if a property is unique to a system, then it cannot be learned.[5] Fodor proposes that learning takes place by hypothesis formation and confirmation. The learner can only form hypotheses using the conceptual apparatus that s/he has in hand at that point. It is therefore impossible for the learner to progress using hypothesis formation and confirmation from a stage in which s/he works with a certain repertoire of concepts to a subsequent stage in which s/he works with those concepts plus some additional property or properties not derivable from the concepts available at the first stage. Fodor's example is that it would not be possible to learn quantificational logic – a system using operators and variables – by hypothesis formation and confirmation based on knowledge of non-quantificational logic. The properties of quantification are not completely expressible in terms of the properties of non-quantificational logic. In general, Fodor's argument goes, it will be impossible to learn a 'more powerful' system.

Does human language exhibit properties that are unique to language, as required by Fodor's argument? The list of linguistic constructs that are – as far as the evidence goes – unique and non-derivable is a long one. Consider for example the facts in (6a–d):

6 a Who did John say that Sue kissed [e]?
b Who did John say Sue kissed [e]?
c *Who did John say that [e] kissed Sue?
d Who did John say [e] kissed Sue?

When an object is questioned (6a–b), the complementiser *that* is optional. When the subject is questioned (6c–d) the complementiser must be absent. There is considerable debate in the linguistic literature concerning the basis for the data in (6a–d), and dialect variation concerning the ungrammaticality of (6c); but this debate is purely in linguistic terms – no one proposes that the constraint exemplified by (6c) derives from more general properties of cognition.

### 6.6.3 CONNECTIONISM AND NEURAL NETWORKS

Frequently, connectionist models share with usage-based models the (implicit or explicit) rejection of Universal Grammar as a restriction on the child's hypotheses. The claim is broadly that what has been described as rules of grammar are an effect of generalisations attributable to, for example, the frequency of forms in the language. Such models attempt to simulate the child's learning using computer trials; the model rewards correct analyses, building up a preference and culminating in the elimination of erroneous analyses (for one introduction to connectionist modelling see Elman et al. 1996; see also Pater 2019 for a succinct summary of the major issues).

Several points can be made with respect to connectionist systems, as represented by work such as Elman et al.'s. There is an extensive literature on the acquisition of the English regular and irregular past tense, beginning with Rumelhart and McClelland (1986), and reviewed in Pinker and Ullman (2002). Pinker and Ullman point out many facts that cannot be attributed to the distribution of data in the input children are exposed to, contrary to connectionist assumptions; for example, children's tendency to over-regularise (to use for instance, *bringed* for *brought*) is not determined by the rate of regular verbs used by parents. Connectionist approaches to the past tense have tended to smuggle in conditions that in effect amount to rules, and the types of phenomena modelled in connectionist approaches is very limited:

> … modelers repeatedly build in or presuppose surrogates for the linguistic phenomena they claim to eschew, such as lexical items, morphological structure and concatenation operations. We predict that the need for structured representations and combinatorial operations would assert itself even more strongly if modelers included phenomena that are currently ignored in current simulations, such as syntax and its interaction with inflection, the massively productive combinatorial inflection of polysynthetic languages. … (Pinker and Ullman 2002: 462)

In addition to the limitation of the range of phenomena currently modelled in connectionism (and this is more or less as true today as it was when Pinker and Ullman wrote their article), such models are capable of learning languages with structures not attested in human languages; for example, a connectionist model could learn a rule that forms questions by inverting the first and second words in a sentence, yet such violations of structure dependency go against a fundamental principle of grammar.

It is important to recognise that there is a rich history behind computer simulations, reviewed by Pater (2019). Pater notes the foundational work of Frank

Rosenblatt (for example, Rosenblatt 1962) on neural networks – models that attempt to simulate mental mechanisms for learning. If such simulations require a rich internal structure for them to work, there is no conflict between the notion of UG and computer simulation.

## 6.7 CONTINUITY VS. MATURATION

The first section of this chapter summarised some general arguments in favour of the view that language development is a biologically guided behaviour. However, a strong component of innate linguistic knowledge does not entail that all constraints and principles of Universal Grammar are available to the child from the outset. It is logically possible that some grammatical properties are programmed to emerge (under appropriate conditions of external stimulation) only after a period in which they are not present in the child's system. It is not difficult to think of analogies with other aspects of human development; for example, the physiological changes of puberty are biologically programmed changes whose exact timing will depend on conditions in the individual's environment, such as diet.

Weissenborn et al. (1992) sketch three broad positions concerning when principles of grammar emerge, roughly as follows:

1. *Strong continuity.* All principles and constructs of Universal Grammar are available at the outset and each grammar formed by the child is a correct (partial) grammar for the language to which the child is exposed.
2. *Weak continuity.* All principles and constructs of Universal Grammar are available at the outset and all child grammars will be possible human languages, in the sense of falling within the patterns of adult grammars (either observed or permitted under the theory). The child's grammar may, however, deviate from that of the language that s/he will ultimately acquire.
3. *Discontinuity.* Some properties of Universal Grammar mature. That is, some properties are biologically programmed to emerge only after a certain period of development. If such a property is an absolute universal (i.e. holds obligatorily for structures and rules to which it is relevant) then child grammars may of necessity fall outside the range of possible human languages.

Since the 1970s, probably the most popular position in the literature has been some version of weak continuity. That position is appealing, allowing for the child to form rules that are not present in the language s/he hears, but nonetheless placing limits on how far s/he can innovate. However, maturational analyses have been proposed, beginning with Felix (1987) and Borer and Wexler (1987) and continuing today with researchers such as Hirsch et al. (2007). For example, Borer and Wexler proposed that the ability to use operations that involve movement into an empty subject position matures in passive sentences, and Hirsch et al. propose the same for 'raising' sentences (sentences such as (70) in Chapter 4, in which the main clause subject derives from the embedded subject). It is very difficult to find clear criteria for maturation. Hirsch et al. suggest the abruptness of change from one (incorrect) analysis to the next correct analysis argues for maturation, but such abruptness could be jogged by the existence of inanimate subjects in passive or raising sentences (as suggested by the work of Becker 2014), leading to a realisation that movement into subject position is an active rule in the child's grammar.

It is important to recognise the distinction between maturation as a hypothesis and the emergentist/constructivist views summarised above: maturation is a biological development as opposed to the consequence of experience. Moreover, although maturation has been challenged as an explanation of developments in the grammar of children aged three and older (for example in the work of Demuth et al. 2010 on the passive), it remains a viable hypothesis for very young children (cf. the debate concerning functional categories, pp. 79–80 above).

## 6.8 LANGUAGE DEVELOPMENT AND COGNITIVE DEVELOPMENT

Is language development an entirely separate process from other aspects of intellectual development, or are there aspects of language development that depend on prior intellectual growth? In this section, we look at evidence on both sides of the debate.

### 6.8.1 THE INDEPENDENCE OF LANGUAGE FROM OTHER MENTAL ABILITIES

The question of whether language progressed in step with other aspects of intellectual abilities was posed in the modern study of language acquisition first with respect to Piagetian stages. Jean Piaget made the claim that language builds on, and requires, other aspects of intellectual growth (see, for example, Piaget 1980).[6] Specifically, it was argued that there are cognitive precursors which link specific (non-linguistic) mental operations to the development of linguistic structures. The latter grows out of the former. For example, Sinclair et al. (1971) suggested that there was a connection between the child's mastery of *conservation* and the development of various linguistic skills and structures. Conservation is the term used in Piagetian theory to refer to the ability to recognise that quantity remains stable although it may be differently distributed in space. A conserving child will realise that when a quantity of water is poured from a tall skinny beaker to a short fat beaker, the amount of water does not change although the water level (distance from the bottom of the beaker) does change. Sinclair et al. suggest that conservation may be a prerequisite for the successful acquisition of the passive. The intuitive basis of this is easy to grasp: the water level changes, but conservers know there is the same amount of water, and in active and passive sentences the position of subject and object is reversed, but we know that the thematic roles of the noun phrases remain the same. There are basically two problems with this type of intuitive connection. First, data that support a constructivist position (conservation preceding skill with the passive) may simply be an accident of two independent developments – a basic statistics course teaches you that correlation does not mean causation. Humans generally take their first steps and use their first words at about the same time, but these are separate developments.

Table 6.1 Complex syntax in a person with Williams syndrome: Crystal (age 15)

| | |
|---|---|
| The dog was chased by the bees. | Full passive |
| Then before they climbed over, they saw baby frogs. | Sentence-initial adverbial clause |
| It was a good movie for children to see because it has action and it has comedy and it has lots of different things in there. | Main clause containing infinitival relative clause with object gap; two clauses conjoined to the main clause |

Second, the facts are not always in accord with the predicted order (skill with passive may precede passing conservation tests, see Speidel 1984).

Over the last thirty years, data have accumulated that argue against cognitive, non-linguistic development as a prerequisite for linguistic development. Another source of evidence for innately given linguistic ability is the language of special populations, including people with Williams syndrome, Down syndrome and other less clearly identified disorders. The common thread in these cases is a linguistic ability that far outstrips what might be expected on the basis of general intellectual ability, diminishing the plausibility of language abilities as some kind of outgrowth of other intellectual skills.

Williams syndrome (caused by a micro-deletion on the long arm of chromosome 7) has been characterised as 'chatterbox syndrome' or 'cocktail party syndrome'. People with Williams syndrome are normally friendly and outgoing, and very talkative. Their scores on standard intelligence tests are well below normal. Yet they have a large vocabulary, including unusual words, and can use complex, well-formed syntax (Bellugi et al. 1993). Examples of the speech of a person with Williams syndrome reported in Bellugi et al. are given in Table 6.1.

Smith and Tsimpli (1995) report the case of Christopher, a disabled person with unclear aetiology. Christopher was in his thirties at the time that Smith and Tsimpli wrote their report. He is unable to look after himself (and hence lives in an institution), and scores very poorly on intelligence tests and other tests of cognitive ability. Yet he has a knowledge of many languages other than English. Smith and Tsimpli (1995: 12–17) list fourteen languages that Christopher can translate, with great success in some instances.

Down syndrome is frequently cited as a case of impoverished language. Although vocabulary is a strong point, skill in morphosyntax is often lacking. But there are cases of considerable linguistic success. In one study, we looked at twenty-five persons with Down syndrome who can read (Goodluck et al. ms.). These individuals vary considerably in their ability to comprehend spoken and written language, but a few are clearly in command of a range of syntactic skills not normally associated with people with Down syndrome. Table 6.2 gives a sample of the productions by Marc, a thirty-two-year-old male, telling a story from a wordless picture book. Marc's sentence production is clearly not fully normal adult usage – it isn't always apparent whether he is using *when/while* as a true adverbial clause marker, or as a way of saying *and then*, and his use of personal pronouns is not always consistent (on occasion he shifts from *its* to refer to a human antecedent to the correct *his* on a later

Table 6.2 Complex syntax in the speech of a person with Down syndrome: Mark (age 32)

| | |
|---|---|
| I see a boy walking a dog. | *-ing* complement to V or reduced relative |
| While he walks the dog … | Tensed adverbial clause |
| … and then he met a frog | *and* conjoined clause |
| … with the net with the fish that he finally caught | Complementiser introduced relative |
| … had a bad day but he decided to go on the leaf | *but* conjoined clause with an infinitival complement to V |
| … get what he wanted | Free relative clause |
| he realized that he's not that depressed any more | Tensed complement to V |
| and the frog was laughing like he was taking LSD | Comparative clause |
| and the boy says "well, no problem in there …" | Embedded direct speech |

usage).[7] But nonetheless Marc is fluent and in command of complex syntax; this is confirmed by a range of structures he understands and produces not included in the samples quoted in Table 6.2. Other examples of exceptional ability among persons with Down syndrome are reported in Segoe (1965) and Rondal (1995).

### 6.8.2 SYNTACTIC BOOTSTRAPPING, THEORY OF MIND AND LANGUAGE GROWTH

Syntactic bootstrapping (not disjunctive from semantic bootstrapping (pp. 73–74), but plausibly a later phenomenon) refers to the acquisition of verb meanings from the syntactic environment in which a verb occurs (for some representative examples of this research, see Landau and Gleitman 1985; Gleitman 1990; Naigles 1990; papers in Gleitman and Landau 1994; Lidz et al. 2003; see also Braine et al. 1990 for a slightly different perspective). As an example, let's take the results of Naigles (1990). Naigles presented video-recorded scenarios with nonsense verbs to children aged 1;11 to 2;3. Two scenarios were used for each verb, one compatible with a transitive use of the verb and the other with an intransitive use; for example, *The duck is gorping the rabbit* and *The rabbit and the duck are gorping* were accompanied by videos in which the duck forced the rabbit into a bending position (transitive interpretation) and in which the duck and rabbit each stretched out an arm (intransitive interpretation). After training on one or other of the sentences (accompanied by both videos), the child was asked, now faced with the two videos separately, *Where's the gorping now? Find gorping!* Children looked significantly longer at the video corresponding to the transitive structure when they were presented with the transitive sentence in the training, and to the video corresponding to the intransitive sentence when they were presented with the intransitive action in the training, indicating that they had processed the connection between the syntactic structure and the action in the video.

Learning words such as *think*, *believe* and *hope* is a significant challenge to the child. As Papafragou et al. (2007) point out, there isn't a direct link between the environment and such words. We can see someone hitting something, and we can extrapolate that the word *hit* represents the action. No such direct evidence is available for belief-type verbs. Papafragou et al. argue that, among the clues to the acquisition of belief verbs, the syntax of the language is important: belief verbs occur with an embedded complement sentence (as, for example, in *Fritz believed that Harry was crazy*). Papafragou and her colleagues found that three- to five-year-old children benefited from the presentation of a complement sentence with a nonsense predicate (*Matt GORPS that his grandmother is under the covers!*); children guessed that the meaning of *gorp* was a belief-type verb more accurately when a sentence with a complement accompanied a videotaped story, than when the video was presented alone. Papafragou et al. argue that the presence of syntax (complement structure) is one of several important clues to the meaning of a word.

Recent research has developed an unexpected version of the relation between cognition in non-linguistic domains and language growth. The twist is that language development may lead to cognitive development, not the other way around. Theory of Mind (TM) refers to our ability to predict the behaviour of others based on our belief of their mental state, i.e. their knowledge, expectations and desires (see, for example, Roeper 2007, Chapter 13). A classic TM test is known as the Sally-Ann test. Sally hides an object, say a key, in a box in the presence of Ann. Ann then leaves the room and Sally moves the key to another hiding place. Ann re-enters the room. Where will Ann look for the key? An adult will answer that she will look in the box, its original hiding place, because the adult has knowledge that Ann is not aware of its new location: an adult has knowledge of the state of Ann's mind. Up until around age four, children will answer that Ann will look in the new hiding place.

De Villiers (2007) and de Villiers and de Villiers (2014) summarise the research arguing in favour of the hypothesis that knowledge of the complement structures of verbs such as *think* and *know* triggers, and is a prerequisite for, passing tests of the TM. De Villiers and de Villiers (2014) write:

> We have argued that the reason for this close affinity between the tasks is that the complement structure permits the language user to represent in a transparent way the content of someone's mind and to differentiate it from reality, so as to judge its truth or falsity ... For example,
>
> John thought *that was his sandwich*, but in fact it was a sponge.
>
> In this example, the complement clause in italics represents a false proposition. *We* know that it is not his sandwich! ... The circumstance of having a false piece in a true sentence is a new occurrence for the child.

Thus, learning the complement structure of verbs lays the ground for development of TM.

Could it be objected (as we did with tests of Piagetian stages and linguistic development) that correlation does not amount to causation. The answer is no. First, there were cases in which the predicted order deviated from the reality in the case of Piagetian stages; such deviation has not been found in the case of TM tests (but see Exercise 3 at the end of this chapter). Second, although correlation is not to be equated with causation in any simple-minded way, if enough cases of x preceding y are found, and no contrary cases are there, we can build an argument for a connection between the two phenomena.

Table 6.3 Dative errors in the speech of children

| |
|---|
| You finished me up a lot of rings. |
| Ursula, fix me a tiger. |
| Jay said me no. |
| Shall at whisper you something. |
| Pick me up all those thing. |
| You put me just bread and butter. |
| Choose me the ones that I can have. |
| Mattia demonstrated me that yesterday. |

*Source:* From Yang (2016), based on Gropen et al. (1989) and Bowerman and Croft (2008).

## 6.9 ERRORS AND ERROR CORRECTION

The subset principle was designed to prevent children from making errors. But they do make errors. If the child is hypothesised to be subject to maturation of grammatical principles, then a principle may mature that leads to the correction of the error, but the principles and structures that potentially mature do not include sensitivity to individual vocabulary items. But data in the input may also serve a corrective purpose. This solution is pursued in Yang (2016), with respect to errors of subcategorisation.

When a child says *Choose me the ones that I can have*, s/he is deviating from the pattern for the verb *choose* (which requires a prepositional form *choose the ones for me*) and must somehow retreat from her/his error to achieve full, adult-like competence. Such errors occur with some regularity in children's speech. Examples are given in Table 6.3, in which children have uttered a double object (NP NP) form for a dative verb that permits only an NP PP frame.

The problem of errors with lexically governed phenomena, and how to correct them, has taxed theorists working within the generative framework for many years (for an early statement of the conundrum see Baker 1979, and for discussions of how to deal with the problem see, for example, Bowerman 1987; Pinker 1989; Randall 1992).

Yang reviews problems with the concept of 'indirect negative evidence', viz. that 'the failure to observe certain forms in the input implies that such forms are ungrammatical' and proposes an alternative mechanism, the *sufficiency principle*. The sufficiency principle – mathematically formulated – adjudicates whether or not there are sufficient lexical items in the vocabulary to merit forming a rule, as opposed to leaving a list of statements in the lexicon.

## CHAPTER SUMMARY

This chapter has defended the innate knowledge of Universal Grammar. This knowledge is supported by the existence of a critical age for normal language acquisition, by the invention of language in deaf populations, and by the linguistic prowess of individuals with genetic defects that lead to below normal intelligence levels. The fact that such knowledge cannot be learned follows from Fodor's argument that it is impossible to learn a system with primitives that cannot be derived from prior existing elements. This does not mean that language and more general cognition are unrelated; indeed, in some cases such as the subcategorisation of belief verbs, language may drive cognitive growth in terms of the development of Theory of Mind skills. Nor does it mean that input is of no importance – we have to learn the particulars of the language we are exposed to, whether it is a matter of general configurations, or details such as whether a rule is justified, rather than a listing of individual statements.

## FURTHER READING

Lasnik and Lidz (2017) provide a very clear summary of issues concerning the quality and quantity of data that the learner is faced with. Marcus (2004) reviews the 'nature vs. nurture' debate.

## QUESTIONS AND EXERCISES

1. Does the fact that Genie stopped before she mastered function words argue in favour of the maturational approach to language learning?
2. People with Williams syndrome develop language later than typically developing children, and perform worse on tasks with age-matched, typically developing children (see Grant et al. 2002). Does this affect the argument made in the text in favour of an innate knowledge of UG?
3. The performance of Christopher (the subject of Smith and Tsimpli's 1995 study) was patchy on Theory of Mind tests. Christopher failed the Sally-Ann test, but did pass the 'Smarties' test. Shown a Smarties tube, and asked what it contained, he replied 'Smarties'. He was then shown that it contained plastic balls. Asked what a friend would think the tube contained, his answer was Smarties, not plastic balls. Smith and Tsimpli report (appendix II) several examples of Christopher's rejection of belief-type verbs (with his suggestions for changes), for example:

   Himself believes John to be happy (Rejected)
   He believes John to be happy (Christopher's correction)
   Do you remember the last time we meet (Rejected)
   Do you remember the last time we met (Christopher's correction)
   John is not as tall as I thought he is (Rejected)
   John is not as tall as I thought he was (Christopher's correction)

What did John say that Mary thought that Peter had misunderstood (Rejected)
What did John say Mary thought Peter had misunderstood (Christopher's correction)

What is your judgement of Christopher's overall performance? Does his performance threaten the idea that knowledge of complement structures is a precursor to development of Theory of Mind?

4. We have concentrated in this chapter on cases in which the linguistic abilities of individuals exceed what we would predict on the basis of other aspects of cognitive ability. The opposite situation does occur. In Specific Language Impairment (SLI), non-linguistic abilities are thought to be unimpaired, whereas parts of the linguistic system (frequently morphological abilities such as tense marking) are faulty (see, for example, Rice et al. 2004; Rice 2017). Does the existence of SLI bolster the argument for the independence of linguistic abilities from other aspects of cognition?

## NOTES

1. A third type of argument comes from the uniqueness of what is acquired; see section 6.6.
2. The principle has been given different names and has been interpreted in different ways; see Clark (1987), Markman (1989) and Wexler and Culicover (1980).
3. This is not to say that child-directed speech does not play an important role in a child's development. The speech to young children is characterised by special acoustic properties (e.g. higher pitch), which may make it a particularly attractive signal for the infant (see Jusczyk and Bertoncini 1988 for a review). And at later stages, speech to children may provide a friendly environment to practise language skills (e.g. rewording or expanding the child's utterance has been argued to enhance linguistic abilities; Demetras et al. 1986).
4. Recent work in phonology has used also the term 'emergent' for the growth of phonetic/phonological features (Mielke 2008; Becker and Tessier 2011); the issue of whether such features are acquired through physiological development is separate from the issue of the development of other aspects of grammar.
5. Fodor made his argument in the context of a debate concerning whether linguistic abilities could 'grow out of' cognitive stages defined in Piagetian theory (section 6.8.1). However, it seems to me that his argument applies equally to usage-based theories of language acquisition.
6. Piaget used the term *constructivism* to describe the prerequisite of other mental abilities for grammatical growth; constructivism has also been used in a more general way in recent research (see section 6.6, usage-based approaches).
7. Marc's mother was a native speaker of Hungarian, and Marc also was fairly fluent in that language (according to the report of his carer; Marc passed away in 2009). His errors with pronouns may have been a result of his knowledge of Hungarian; it is common for Hungarian speakers who have learned English as a second language to substitute *its* for a gendered pronoun.

## CHAPTER 7

# PERFORMANCE DEVELOPMENT

It is vital to the study of child language that we understand if a child's non-adult behaviour is a matter of immature grammatical rules or an immature system for processing and producing sentences. In this chapter, we look at the nature of the adult sentence processing and production devices and attempt to estimate the degree of similarity to the child's device. This will lead us to in part re-evaluate hypotheses about the learning mechanism that the child uses and other proposals concerning adult vs. child performance mechanisms.

### 7.1 ESTIMATING COMPETENCE

Linguists use grammaticality judgements by native speakers of a language as the basis for the theories they propose. Although there has been recent controversy about such reliance on native speaker judgements, the practice has nonetheless been validated (Sprouse and Almeida 2012). By and large, children are not able to give the types of overt grammaticality judgement that adult native speakers give, particularly in the case of ambiguous sentences and/or subtle distinctions (see Gordon 1996 for one assessment of a child's ability to give judgements). Estimates of children's knowledge must to a degree rely on the interpretation of observational and experimental data. Because of the indirect nature of the evidence, it is easy to overestimate or underestimate what children know – their grammatical competence.

It is almost certainly true that it is more usual to underestimate a child's competence than to overestimate it. Spontaneous production may be a poor guide to competence because the child may not produce examples of all the sentence types that are within her/his competence. And tests of comprehension may underestimate children's abilities for a variety of reasons, including difficulty of the task and the complexity of the sentences confronting the child. An example involving relative clauses will illustrate how experimental tests can produce results that make it look as if children know less than they do.

Relative clauses have been observed in children's spontaneous speech in the third year (see for example, Limber 1973), and by age three children can be induced to produce relatives in experimental situations (see, for example, Hamburger and Crain 1982 and Labelle 1990), although children's productions are not always true to the adult grammar. Early comprehension studies showed that preschool children did rather poorly in interpreting some types of relative clauses. Using an act-out task, Sheldon (1974) and Tavakolian (1981) found that three- to five-year-old children frequently misconstrue a relative that modifies a direct object as referring to the subject of the sentence. Thus, a sentence such as (1) will be interpreted as referring to mean that the horse, not the cow, pushes the dog:

1 The horse kicks the cow that pushes the dog

Tavakolian argued that the child misconstrued the relative clause, which should be embedded under the object NP, as a conjoined clause, to yield the interpretation:

2 The horse kicks the cow and pushes the dog

However, children's performance with sentences such as (2) can be altered by manipulation of the internal content of the relative. Also in an act-out experiment, Goodluck and Tavakolian (1982) found that if the relative verb is intransitive (3a), or contains an inanimate object (3b) rather than an animate object (3c), the correct interpretation increases, as shown by the percentages correct after each example in (3a–c):

3 a The horse kicks the cow that jumps up and down 76% correct
b The horse kicks the cow that knocks over the table 69% correct
c The horse kicks the cow that pushes the dog (=1) 49% correct

Such results argue that the performance mechanism has a vital role to play in children's success. Intransitive relatives may be easiest because there is no object to process, and inanimate objects may be easier because of a general expectation that objects are inanimate.

We look now at some of the properties of adult and child mechanisms for comprehension and production of sentences, returning to debates concerning immature grammars vs. immature performance mechanisms at the end of the chapter (section 7.5).

## 7.2 THE ADULT PROCESSING MECHANISM

The picture of the adult processing device we will present here is based on many experimental findings but nonetheless has many open questions about it. For useful reviews of the literature, see Fodor (1995), Tanenhaus and Trueswell (1995) and Treiman et al. (2003). Here are two facts that are solidly grounded: the mechanism is quick and efficient, and its structure is revealed by the errors it makes.

### 7.2.1 GARDEN PATHS AND ERRORS

In listening to a sentence, we effortlessly construct a meaning. A 'sensible' first step in the processing procedure is to look up the words in the mental dictionary and consult our knowledge of grammar for a syntactic analysis. Some sentences present no trouble. But other sentences trip the processor up. The most famous example in the literature is:

4 The horse raced past the barn fell

A naive listener or reader of (4) (one who is not already familiar with the example) finds it impossible to fit in the final word *fell* with the analysis already computed. The correct syntactic analysis contains a reduced relative clause modifying the subject, yielding a meaning equivalent to (5):

5 The horse which was raced past the barn fell

When the pragmatics are right, as in (6), we can do without the extra words in (5), but otherwise we inexorably analyse the word *raced* as the main verb, not the verb in a reduced relative clause:

6 The horse trained by an expert fell

The misanalysis of (4) is accounted for by a principle of sentence processing: build the minimal structure compatible with the input, which yields the structure (7) rather than the correct structure (8):

7

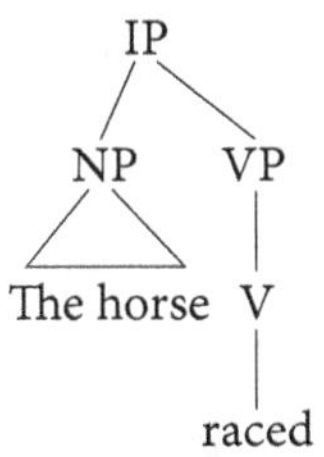


8

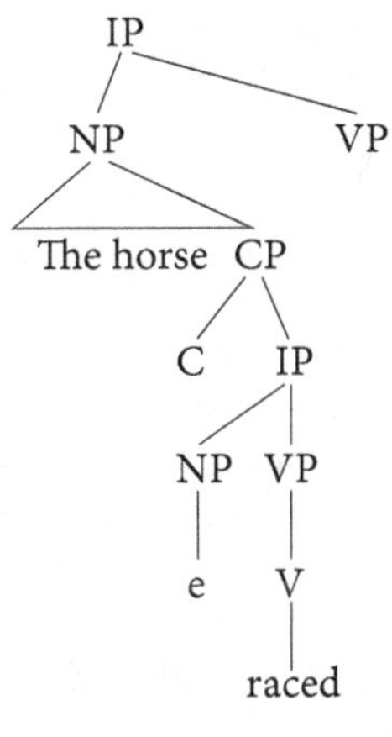


We get the clue to the correct analysis for (6) by the fact that horses do not generally train things, whereas the incorrect analysis is promoted by the fact that horses do race. Sentences such as (4) are referred to as *garden-path sentences*, since they lead the hearer/reader up the metaphorical garden path of a wrong analysis.

A further type of garden-path sentence arises from the processor's inclination to chunk together the last-analysed string with new input (Frazier and Rayner 1982). In (9), the phrase *the king* and the phrase *his horse* are erroneously construed as subject and object of the subordinate verb, when in fact the subordinate verb is intransitive and *the horse* is subject of the main verb:

9 When the king rides his horse is always groomed

(A comma between *rides* and *his horse* disambiguates the sentence, but this clue is not available in speech and a pause may not be sufficient to prevent the error.)

The sentence or clause also forms a point at which the processing device makes a break in the input utterance. There is interaction between what determines that a clause is complete and the information contained in the mental lexicon. The subcategorisations of an individual verb allow for embedded complement sentences, which extend the domain that the processor works on. And length of constituents may enforce a break before a clause is completed, i.e. the processor may divide the sentence because it contained, for example, a very long noun phrase.

Another principle of sentence processing dictates that we resolve question formation and other discontinuous structures as soon as possible. In a question such as (10), there is only one place for the base position of the question word *what* (as object of *with*) but nonetheless the sentence processor leaps to the wrong analysis and construes *what* as object of *eat*:

10 What did Sarah eat a banana with?

For example, many experiments have shown that in a question such as (10), the processing device slows down on encountering the words *a banana*, consistent with the device having *what* as object of *eat*, only to discover its error when the sentence continues. This has been dubbed the result of the *active filler strategy* (Frazier and Flores d'Arcais 1989): the filler word (*what*) is slotted into the sentence at the first available opportunity.

### 7.2.2 A ROUGH SKETCH OF THE ADULT PROCESSOR

The findings summarised above can be accommodated within several types of model. The approximate model adopted here is sketched in Figure 7.1. Some of the properties of the model in Figure 7.1 seem self-evident: the processor has to look up the meaning of words, which are in the lexicon. And you can't begin working out the structure until you have accessed the syntactic category of words, which is also part of the lexical entry (although, as we see below, it is possible that we have pre-packaged formulas into which words of the right category are fitted).[1]

There are other properties of the proposed processor that are less obvious. We have assigned no place for real-world knowledge, yet evidently it must figure into the processing procedure, as we saw with the contrast between (4) (*The horse raced past the barn fell*) and (6) (*The horse trained by an expert fell*). And so far we have made no mention of procedures for resolution of remaining ambiguity, ambiguity left after the phrase structure has been resolved. In a sentence such as (11), the pronoun *he* may refer to either *Fred* or *Ed*, but generally an adult prefers the interpretation in which the pronoun refers to *Fred*:

11 Fred saluted Ed when he entered the building

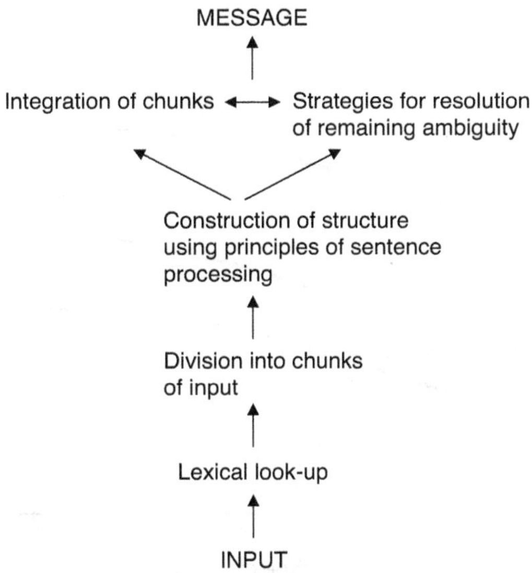


*Figure 7.1 Structure of the adult sentence processing mechanism*

## 7.3 CHILDREN'S PROCESSING

The general hypothesis that we will pursue in this section is that sentence processing by children proceeds in the same way as sentence processing by adults, but there are several differences, which may be due to a deficiency in the child's vocabulary, in her/his grammatical rule store, or a lesser quantitative ability to process (as the result of a reduced memory capacity and/or a lesser ability to recover from errors).

### 7.3.1 AN OVERALL SIMILARITY

Early studies demonstrated a broad similarity between child and adult processing. Tyler and Marslen-Wilson (1981) asked children – aged five, seven and ten – and adults to monitor a pre-designated word in an orally presented prose passage. In the monitoring task, the subject must press a button as soon as s/he hears the word that s/he is told to listen for; in Tyler and Marslen-Wilson's experiment, the word was specified before the passage began. There were three types of passage in the experiment:

| | |
|---|---|
| *Normal prose*: | A sequence of grammatical sentences with a sensible meaning. |
| Example: | John had to go back home. He had fallen out of the swing and had hurt his *hand* on the ground. |

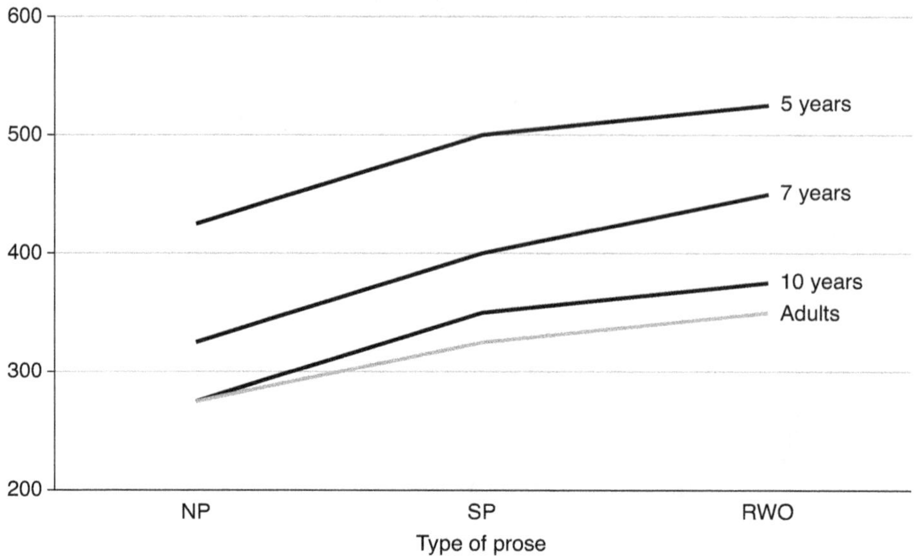


*Figure 7.2 Mean reaction times (msecs) in identical monitoring for each prose type*
*Note:* NP = normal prose; SP = syntactic prose; RWO = random word order.
*Source:* Adapted from Tyler and Marslen Wilson (1981: figure 1).

| | |
|---|---|
| *Syntactic prose*: | A sequence of syntactically correct sentences, in which the choice of vocabulary items permits no coherent semantic interpretation. |
| Example: | John had to sit on the shop. He had lived out of the kitchen and he had enjoyed his *hand* in the mud. |
| *Random word order*: | A sequence of unstructured words, divided into sentence-length chunks corresponding to the normal prose and syntactic prose sentences. |
| Example: | The on sit shop to had John. He lived had and kitchen the out of his of had enjoyed *hand* mud in the. |

(These examples are Tyler and Marslen-Wilson's translations from the Dutch originals.) The word to be monitored in the examples is *hand*.

The monitoring task, because it required the subject to respond as soon as s/he recognises the word s/he is searching for, provides a measure of processing as the sentence is input. Figure 7.2 gives the mean millisecond reaction times in Tyler and Marslen-Wilson's experiment. The results support in a general way the idea that the child uses a sentence processor with the same structure as the adult's. The overall pattern of results is similar to that for adults, although the children's reaction times are slower.

*Figure 7.3 Organisation of visual scenario*

*Source:* Trueswell et al. (1999).

### 7.3.2 CHILDREN'S (IN)ABILITY TO RECOVER FROM ERRORS

Children may not merely be slower to respond than adults, they may erroneously jump to a conclusion which they cannot recover from. This was demonstrated by Trueswell et al. (1999). The verb *put* in English requires both a direct object and a locational phrase:

12 Put the frog on the napkin
13 *Put the frog
14 *Put on the napkin (where *on the napkin* is construed as a phrase, and not the object of *put on*, meaning *wear*)

Trueswell et al. gave orders such as (15) to five-year-old children in the context of a visual display in which there were two frogs, one on a napkin and the other not, a napkin with nothing on it, and an empty box (see Figure 7.3):

15 Put the frog on the napkin in the box

The grammar requires that *in the box* is construed as the location where the frog ends up, and *on the napkin* is construed as the right frog to choose to be put there. The task was acting out, with recording of the child's eye movements as s/he scanned the scene.

The results showed that five-year-olds persistently made the error of placing the frog that was not on the napkin onto the napkin, i.e. they construed *on the napkin* as the location to which the frog must go, not as a clue to which frog was to be chosen. Adult participants did not make the same mistake. Such errors on the part of children were not made when the order was made unambiguous by clear use of a relative clause structure:

16 Put the frog that's on the napkin in the box

Thus, the children jumped at the first possible analysis before they heard the end of the instruction. Their lack of revision indicates that their initial misanalysis was so entrenched that it could not be corrected.

### 7.3.3 THE ACTIVE FILLER STRATEGY IN CHILDREN

Cross-linguistic studies have argued that the active filler strategy guides the child's processing of questions. Omaki et al. (2014) investigated globally ambiguous questions in English and Japanese. In questions such as those in (17), the question word can be interpreted as located in the main clause or the subordinate clause:

17 a Where did Lizzie tell someone that she was gonna catch butterflies?
   b Doko-de Yukiko-chan-wa choucho-o tsukamaeru-to itteta-no?
   Where-at Yukiko-Dim-Top pro butterfly-Acc catch-Comp was telling-Q
   'Where was Yukiko telling someone that she will catch butterflies?'
   (Acc = accusative case marker; Comp = complementiser; Dim = diminutive marker; Q = question marker; Top = topic marker)

The order of clauses differs in the two languages: main clause followed by subordinate clause (English) vs. subordinate clause followed by main clause (Japanese). Yet the question word can refer to either clause.

Questions such as (17a–b) were asked following a story that supported both interpretations of the ambiguous question (following the methodology in de Villiers et al. 1990 and de Villiers and Roeper 1996). In each language the participants were an average age of approximately five-and-a-half years. Omaki et al. found a robust preference for the first location for the *where/doko-de* phrase – i.e. the main clause in English, but the subordinate clause in Japanese.[2] This demonstrates that the active filler strategy overrode the difference in syntactic configuration for the two languages.

Goodluck et al. (1992) used the story-after-question technique to test three- to four-year-old children's sensitivity to the block on movement from an adverbial clause, illustrated in (18):

18 *What did John read Dickens before writing?

When the question contained two optionally transitive verbs, as in (19),

19 What did the fox eat before whistling?

they found a very strong preference to construe the question word as object of the main verb (*eat*), as opposed to the subordinate verb (*whistle*). A similar preference was not found for questions such as (20), also with two optionally transitive verbs,

20 Who did the zebra ask to kiss?

indicating that the children were capable of responding with an answer that referred to the lower clause object position. Goodluck et al. interpreted these results as indicating that children were aware of the constraint illustrated by (18).

However, this was not an interpretation that would stand up to cross-linguistic evidence. Goodluck et al. (1995) compared the processing of questions in English to the processing of questions in Akan, a Kwa language spoken in Ghana. Akan, unlike English, freely allows questions equivalent to (18):

21 Dɛn na Ama kanee Graphic ansa na ɔrekyerɛw?
What FOC Ama read Graphic before she:wrote
'What did Ama read the Graphic before she wrote?'
(FOC = focus marker)

Saah and Goodluck (1995) proposed that adult Akan uses a pronominal reference mechanism, rather than movement, to form questions, and the language is not subject to the constraints attendant on a movement mechanism, including the block on reference between a question phrase and a position inside an adverbial clause. (21) is fully grammatical in Akan.[3]

Goodluck et al. translated the materials in the earlier study (Goodluck et al. 1992) into Akan and tested five- to six-year-old Akan-speaking children in Ghana. The children behaved very similarly to the English-speaking children, avoiding reference into an adverbial clause, but allowing reference into a complement to the VP. When Akan adults were tested on the same test (Saah and Goodluck 1995), the same pattern was found. In short, both English-speaking children and adults and Akan-speaking children and adults avoided co-reference between a question phrase and a position inside an adverbial clause, and both allowed reference inside a complement to a VP.

What can be made of these results? The first possibility is that we have made an error in the analysis of adult Akan. This was ruled out by Saah and Goodluck, who also tested Akan adult speakers on a reflective grammaticality judgement task and found acceptance for questions such as (21), contrary to English speakers, who rejected questions such as (18). Saah and Goodluck proposed a distinction between rapid, non-reflective tasks, such as question response, and graded grammaticality judgement tasks; the former may give a snapshot of the sentence processing procedure (see also Goodluck et al. 2017). If this is the case, then the results argue for the operation of the active filler strategy in both child and adult processing, and do not necessarily indicate sensitivity to the constraint in English.[4]

### 7.3.4 PROCESSING EFFECTS OR ACQUISITION PRINCIPLES?

One important question that has largely been unasked in the literature concerns whether putative principles of language acquisition are in fact the effects of the language processing device. Consider the subset principle (Chapter 6, sections 6.4 and 6.9). The subset principle prevents error on behalf of the child by adopting the most restrictive grammar available given the evidence at hand. However, we noted in Chapter 6 that the subset principle ran into problems with data that went against its predictions; in addition, there are concerns about its psychological plausibility (does the child really actively compare possible grammars and choose the least liberal one?). The results from Akan just summarised argue that in some cases the effect of the subset principle may be mimicked by the sentence processing device. Both English and Akan permit reference of a question word into a main clause, but only Akan permits reference into adverbial clauses and other clause types from which the operation of movement is banned in languages such as English. The active filler strategy dictates that only the choice of main clause extraction is made, in effect replicating – and doing away with the need for – the subset principle.

### 7.3.5 RESOLUTION STRATEGIES AND DISCOURSE STRUCTURE

In sentences such as (11) above, repeated here, adults generally interpret the main clause subject, and not the object, as the referent of the pronoun *he*, even though either interpretation is possible under the rules of grammar (see the binding theory, section 4.4.1):

11 Fred saluted Ed when he entered the building

Children do not share this preference. Many studies have shown that when the referent of a pronoun is ambiguous, children just guess with respect to the antecedent. For example, Arnold et al. (2007) tested three- to five-year-old children on two sentence discourses such as the following, in which *Puppy* and *Panda Bear* were clearly established as male entities by the experimental set-up:

22 Puppy is having lunch with Panda Bear. He wants some milk.

The child's task was to move the mentioned object (milk in the example) to the front of either *Puppy* or *Panda Bear*. Even the oldest children performed at chance on the choice of which animal the object was moved to, in contrast to adult subjects, who chose the first mentioned animal (*Puppy*) in 88 per cent of their responses. Although studies such as Hickman and Hendricks (1999) and Tsimpli et al. (2014) reveal cross-linguistic differences rooted in the types of anaphor (overt or null) and the structure of the language, the basic finding that reference to the first mentioned entity is not acquired until into the school years is a robust one.

It may be that children rely on thematic (semantic) roles in a stronger way than adults do. Solan (1983: chapter 5) varied the voice (active vs. passive) of two conjoined clauses, such that the syntax of the conjuncts was the same or different. Examples of his materials are given in (23):

23 a The dog hit the sheep, and then she hit the cow
b The dog was hit by the sheep, and then she was hit by the cow
c The dog hit the sheep, and then she was hit by the cow
d The dog was hit by the sheep, and then she hit the cow

Children aged five to eight years and adults were tested. The participants were informed that all the animals were female (hence the pronoun *she* was used throughout). The experimenter read the sentences and acted out the first conjunct. Participants acted out the second of the two conjuncts. The basic finding was that children made the pronoun refer to *the dog* in sentence types (a) and (b), and to *the sheep* in sentence types (c) and (d), i.e. that they matched the pronoun to its semantic role in the first clause. Adults were more inclined to choose the subject of the first clause for all four sentence types (although they were less consistent for types c and d).

In general, children may be less adept at following or developing interpretive strategies or restrictions that apply at later stages of sentence processing, a finding consistent with the distinctions found between sentences in Ruigendijk et al's (2011) study (Chapter 4, p. 62) and with the results of Janke (2018b), who studied children's interpretations of the non-obligatory control construction in (24):

24 [PRO Rowing the boat clumsily] made Luna seasick

The PRO subject in (24) can refer to *Luna* or to an entity not mentioned in the sentence. Janke found that children aged six–eight years were less likely to follow a cue to sentence-external reference of PRO from sentences preceding (24) (*Ron is going out on the lake; Ron takes the oars awkwardly*) than children aged eight–eleven were.[5]

## 7.4 SENTENCE PRODUCTION

It seems a little anomalous that although data from children's utterances have had such a prominent role in theorising about the development of language, there has been little attention to a comparison between the adult production device and the child's production device. Here we will summarise some facts about adult production and some proposals concerning the child's production mechanism.

### 7.4.1 THE ADULT MECHANISM

The adult production mechanism first conceptually plans an utterance (formulates an idea), then in tandem formulates a syntactic representation and performs lexical look-up. An important planning unit is the clause. The mechanism has a built-in correction device, which means that although some errors creep in, potential mis-speech such as *a pearly period* (for *an early period*) are corrected, with the correct determiner (*a* rather than *an*) used. See Garrett (1980) and Dell (1995) for summaries of the structure of the production mechanism and typical errors that the adult processor makes. A basic finding concerning the adult production mechanism is the preference for organising sentences in the order *given* (elements) before *new*, and *short* (elements) before *long*. These organisation principles are complementary, since given elements such as definite pronouns are also short elements (*he*, *she*, *they*, etc.). However, the two are distinct, as demonstrated by Arnold et al. (2000).

Table 7.1 Number of filled pauses at utterance onset and other locations

| | Beginning only | Other places only | Both |
|---|---|---|---|
| Younger children | 13 | 39 | 5 |
| Older children | 24 | 29 | 3 |
| Adults | 46 | 2 | 3 |

*Source:* McDaniel et al. (2010).

### 7.4.2 CHILDREN'S SENTENCE PRODUCTION

McDaniel et al. (2010) studied sentence planning in children aged three to eight years and adults, measuring their patterns of pausing when experimentally prompted to produce relative clauses. They found a basic similarity between children and adults, both using a silent pause before the complementiser that introduces a relative clause, for example, before *that* in (25):

25 ... the bear that the king is hitting

There were, however, differences between the children and adults. Younger children (aged three to five) differed from adults in using filled pauses (pauses filled with elements such as *um*) in more positions than simply the beginning of the utterance; older children (aged six to eight) showed performance with filled pauses between those of the younger children and adults, as shown in Table 7.1.

McDaniel et al. argue that the overall architecture of the sentence production mechanism is the same for children and adults, but that children and adults differ in the amount of advance planning that they can make. They draw an analogy with crossing a creek by jumping from stone to stone: 'Children and adults land on stones that are positioned in the same places. But adults figure out the path before starting across, whereas children do some of the figuring out on the way' (McDaniel et al. 2010: 92).

The general hypothesis that adult and child production mechanisms are structured in a similar way is enhanced by the similarity of speech errors made, and by the finding that they are influenced in the same way by prior speech. For example, both children and adults are more likely to utter a passive sentence if they have previously heard (have been *primed* by) a passive (Branigan and Messenger 2016).

Moreover, children appear to use something similar to the adult organisational principle of given-before-new. Karmiloff-Smith (1980, 1985) showed that young children organise their discourse in such a way that the topic of the discourse is consistently made subject or utterance-initial element of each sentence and is expressed by a pronoun or an elided (null) NP. Asked to relate the events in a sequence of pictures featuring a small boy and a balloon-seller, one child produced the following discourse (B = boy, referent of the pronoun):

> A little boy is walking along. He (B) walks off in the sunshine. He (B) sees a balloon seller. He (B) wants a green balloon. He (B) gets one. He (B) lets go of the balloon and then he (B) starts crying. (Karmiloff-Smith 1980: 242)

The boy, established as discourse topic of the first sentence, is thus consistently made pronoun subject of subsequent sentences in the discourse. When asked to relate the events in the same pictures presented in isolation, children vary as to choice of the *boy* or the *balloon-seller* as subject of the sentence, showing that the choice of a single entity as pronoun subject is not merely an artefact of (for example) the pictures used as stimuli.

### 7.4.3 THE EARLIEST STAGES

The studies of children's production just summarised deal with children aged three and over. What about younger children? We saw in Chapter 4 that it is characteristic of English-speaking two-year-olds and younger three-year-olds to drop the subject of main clauses, but not of embedded clauses. Rizzi (2008) speculates that the two-year-old infant uses a strategy to cope with the inherent limitations of an immature production device: 'When production begins, the child initially assumes all parametric values which facilitate the task of the immature production system by reducing the computational load, and which are consistent with her current grammatical knowledge' (Rizzi 2008: 38).

Young English-speaking children set the value of the null subject parameter very quickly, blocking subject drop except in special contexts; but subject drop is permitted in certain English dialects, such as that studied by Thrasher (1977: see chapter 4). It is that possibility which Rizzi appeals to, assuming that the computational load is reduced by the absence of an overt subject. This accords with the observation of Bloom (1990), that children's subject drop is more frequent when the verb phrase is longer, requiring more effort from the production device.

## 7.5 SEPARATING IMMATURE GRAMMARS FROM IMMATURE PRODUCTION/PROCESSING MECHANISMS

The previous section (7.4.3) raised the issue of where the borderline between a child's grammar and performance mechanisms is to be drawn. Overall, we saw that children learning English correctly set the parameter for dropping of subjects to the English setting, with the twist of allowing subject drop in circumstances allowed by some English dialects, and they are influenced by computational complexity in permitting the omission of a subject.

This is the familiar dilemma: is the child using a non-adult grammar, or does s/he have an adult-like grammar but is swayed by ease in executing performance mechanisms? A recent debate has centred around such an issue concerning an observed asymmetry in the production and comprehension of questions and relative clauses in which the object rather than the subject is focused, as in (26a–d):

26 a Who kissed the zebra? (subject question)
   b Who did the zebra kiss? (object question)
   c The dog licked the elephant that kissed the zebra? (subject relative clause)
   d The dog licked the elephant that the zebra kissed? (object relative clause)

In general, children have more difficulty with object questions and relatives than subject relatives and questions (though there are exceptions, as we will see below). Two opposing positions have been put forward to explain this asymmetry:

*Position A* (immature grammar): The child imposes a stricter version of an adult restriction used in some languages, which prevents an element moving across another element, when that other element is a target for movement. This line of research developed as an outgrowth of work on adult languages that began with Rizzi (1990). The basic idea is that in child grammar there are intervention effects: effects that, for example, account for the ungrammaticality of questions in adult English such as (27):

27 *How do you wonder who behaved _?

Example (27) is ungrammatical on Rizzi's account because *who* intervenes between *how* and its position as object of *behaved.* The gist of the proposed child version of the grammatical restriction is that *the zebra* in (26b and d) intervenes between *who/the elephant* and its position as object of *kissed.* For details of the technical formulation of the child restriction, see Friedmann et al. (2009).

*Position B* (mature grammar; processing pressure leads to error): The processing account given by researchers such as Kidd et al. (2007) points to factors that ameliorate the difficulty of object relatives (object questions have not been discussed to the same extent). Specifically, Kidd et al. found in their study that children produced and understood object relatives better when the relative had an inanimate head noun and the subject of the relative clause was pronominal. Similar observations have a long track record (see for example, Limber 1976), and are bolstered by researchers such as Gerard et al. (2017).

At present it is not possible to firmly adjudicate between Position A and Position B. Neither position is in principle incompatible with the other, although Position B has the potential to eliminate the need for Position A. Some potentially fruitful lines of enquiry lie in the development of movement operations for relative clauses, since movement is at the heart of Position A and existence of movement has been queried for child relatives (for example, Labelle 1990, 1996; Goodluck and Stojanović 1996; Goodluck et al. 2006; see also Guasti and Shlonsky 1995 and Guasti and Cardinaletti 2003 for contrary views). Moreover, the nature of the question phrase in the putative child grammar affects performance. Goodluck (2010) presents evidence that a less specific question phrase such as *which animal* does not present the same difficulty for children when the child moves over an intervener as a more specific phrase such as *which zebra*, something that is not accounted for by Friedmann et al.'s analysis.

## CHAPTER SUMMARY

We have seen in this chapter evidence that the child's comprehension and production mechanisms are roughly structured in the same way as the adult's, but there are differences. Some of these differences are easily attributed to a lesser processing capacity, such as McDaniel et al.'s (2010) finding that children's filled pauses are more frequent than those of the adult, and Rizzi's (2008) account of Bloom's finding that children drop subjects more often when the VP is longer. Others are not so amenable to such an explanation, such as children's seemingly greater reliance on semantic (thematic) roles in analysing the input, as found in Solan's (1983) study. At the end of the chapter we raised questions about competing explanations (child grammars vs. immature performance systems), questions that remain largely unresolved.

## FURTHER READING

Felser et al. (2003) is a report on six- to seven-year-old children's processing of ambiguous relative clauses, using a self-paced listening task. The article goes beyond the material covered in this chapter by examining the differences between listening span for children and (lack of) sensitivity to lexical structure. It is not a technically demanding read and introduces processing principles that supplement those described here. Clahsen and Felser (2006) contains a review of studies of language processing by children. See Goodluck and Kazanina (2020) for some discussion of performance from the perspective of Chomsky's Minimalist framework.

## QUESTIONS AND EXERCISES

1. Gruber (1967) gives an example of how a child's spontaneous speech can give the impression of being more adult-like than in fact it is. A young two-year-old produced questions such as 'Where went the wheel', meaning 'Where did the wheel go?' Questions of this sort appear to involve inversion of the subject (*the wheel*) and the verb (*went*). But Gruber argues that the child's questions do not involve an adult-like inversion rule. Rather, he suggests that other elements in the child's speech at the period (such as the restriction of subjects to pronouns in utterances other than questions) argue for a child grammar in which the 'subject' in questions such as 'Where went the wheel?' is in fact a sentence final topic.

   Discuss Gruber's analysis in the light of the data on null subjects summarised in Chapter 4.

2. In Chapter 4, we looked at children's knowledge of Principles A and B of the binding theory, and the controversy over the source of the 'Delay of Principle B'

effect. Suppose we ask children to listen to sentences containing reflexive pronouns (subject to Principle A) and definite pronouns (subject to Principle B), faced with an array of three dolls named Fred, Bill and Harry.

a Fred wanted Bill to shave himself
b Fred wanted Bill to shave him

If we measure the child's eye movements over the array of dolls, what would we expect to find?

3. Consider another mind experiment. Children are asked to listen to sentences of the following types and detect a mispronounced word (sentence type (a) has an embedded wh-question and sentence type (b) has an embedded *if*-clause). The word to be mispronounced is italicised in the examples.

   a Tom asked what Sue wrote *letters* to Bill for
   b Tom asked if Sue wrote *letters* to Bill often

   In which sentence type (a or b) would you predict that there would be longer reaction times to identify the mispronounced word? Why?

4. Bever (1970) and Maratsos (1974a) found that at around the turn of the fourth year there was a dip in performance with passive sentences with action verbs, i.e. children got worse at the passive, acting out sentences such as (b) as their active equivalents (a):

   a The lion bit the tiger
   b The lion was bitten by the tiger

   Bever explained this as an increased reliance on a strategy by which an N-V-N sequence was interpreted as subject-verb-object. What alternative explanation could be given in terms of grammatical development?

5. De Villiers and Roeper (1995) told short stories to preschool children, followed by questions such as (a) and (b):

   a How did the man who hurt his leg get home?
   b How did the man rescue the cat who broke her leg?

   In (a), the first verb (*hurt*) is inside a relative clause, and so linking to a position inside the VP is blocked (**How did the man who hurt his leg?*). In (b), the first verb (*rescue*) is the main verb, and so linking to a position inside the VP is permitted (*How did the man rescue the cat?*). De Villiers and Roeper found that children categorically avoided the first VP in (a) as a location for the question word *how*, but they did not avoid the first VP as a location for *how* in questions such as (b). De Villiers and Roeper interpreted this as evidence that children obeyed the constraint on movement that prohibits linking a question word to a position inside a relative clause.

   What alternative explanation can be given for their findings? Hint: how well formed in terms of argument structure (completness of the sentence) are the questions *How did the man who hurt his leg?* and *How did the man rescue the cat?*, independent of the fact that you have linked to a position inside a relative clause vs. a main clause VP?

## NOTES

1. Figure 7.1 should not be interpreted as indicating that there are rigid stages in which, for example, all lexical look-up precedes the division into chunks, which then precedes the application of processing principles. Rather, the model should be interpreted as giving some stages a head start, with activity at all levels potentially taking place simultaneously.
2. Such a preference is modulated in some circumstances; see below.
3. In a paper published after this chapter was in press, Korsah and Murphy (2019) propose that movement is involved in Akan questions; they rely on the presence of a null or resumptive pronoun to legitimise questions such as (21). Frank Tsiwah has provided data that challenges the movement analysis (paper in preparation). The difference between the analyses of Saah and Goodluck (1995) and Korsah and Murphy does not affect the point made below, i.e. that the child takes the first available option to unload the question word.
4. The extraction from the complement to a VP is dealt with by the subcategorisations of the verb 'opening up' the complement material (section 7.1.1 above). See Goodluck et al. (1992) and Goodluck (2002) for a more detailed account of the difference between complements to the verb and adjunct clauses.
5. The idea that operations towards the end of the processing model are less accessible to children runs counter to the idea that children are very sensitive to contextual information, an idea that was popularised by, *inter alia*, Hamburger and Crain (1982). In my opinion, the evidence in favour of such sensitivity is not strong; see Goodluck (1990) and Eisenberg (2002) for some relevant findings.

# APPENDIX 1

## METHODS IN CHILD LANGUAGE RESEARCH

This appendix summarises the main procedures available for child language research. Four edited volumes (McDaniel et al. 1996; Sekerina et al. 2008; Unsworth and Blom 2010; Hoff 2012) provide chapters that give more detailed descriptions of the techniques listed below.

It is important to make the point that there is no one method that stands out as better than another; they each have advantages and disadvantages. In addition, methodology cannot replace hard, critical thinking: about the ideas that drive a study, and into the practicalities of doing the study. The methods vary considerably with respect to their reliance on equipment, the amount of data that can be gathered, ease of administration and suitability for probing different areas of linguistic ability. We begin with spontaneous production as a source of data, and then go on to experimental methods.

Spontaneous Production This method predates tape recording. The researcher (in the earliest studies, the parent) notes or tape-records what the child says and attempts to make an analysis for either the whole grammatical system or particular constructions. The advent of the CHILDES database has opened up the analysis of spontaneous production data to all researchers with access to a computer (MacWhinney 2000). The CHILDES database consists of corpora from quite a wide range of languages. Some of these are edited/tagged for grammatical functions, and there are programmes that allow a researcher to select out particular constructions. The PhonBank consists of corpora that focus on early child phonology.

> Age range for which corpora are available: Various.
> Equipment: A computer with online access.
> Types of linguistic knowledge studied: All areas, but the corpora may give limited access to contextual information that restricts the usefulness for certain analyses.
> Further Reading: Demuth 1996; Stromswold 1996; Corrigan 2012; Lieven and Behrens 2012; Naigles 2012.

The experimental methods available for the study of child language can be divided into 'High task demand' and 'Low task demand' measures, organised alphabetically in the lists below. The former require some degree of active involvement by the child; in the latter, the child reacts to the stimulus unconsciously.

## HIGH TASK DEMAND METHODS

ACT OUT In this task, the experimenter utters a sentence and the child performs actions with toys that indicate what the child thinks the sentence means. A potential difficulty with the task is that it only reveals one possible interpretation of the sentence, not the full range that the child permits.

AGE RANGE FOR WHICH THE TASK IS SUITABLE: 3 years and up.
EQUIPMENT: Dolls/toy animals and other props; video-/audio-recorder (optional, but desirable).
TYPES OF LINGUISTIC KNOWLEDGE STUDIED: Syntactic and semantic knowledge in a wide range of areas; some sentence types (for example, questions) are not easily acted out.
FURTHER READING: Goodluck 1996.

COLOURING BOOK TASK Although children's attempts to draw their understanding of sentences have had a long – if patchy – history, new technology has made the task doable. The child is presented with black and white pictures and is asked to colour in a relevant part. For example, pictures of a girl spraying water on a boy, followed by the girl and the boy each holding an uncoloured apple, may be paired with the sentence *The girl washed the boy before eating the red apple.* The adult grammar rules that it is the girl who eats a red apple, not the boy, and the correct answer is to colour the apple that the girl is holding. Computer presentation makes for quick and unambiguous responses: once one of the apples had been coloured using a touchscreen computer, the programme prevents any further coloring.

AGE RANGE FOR WHICH THE TASK IS SUITABLE: 3 years and up.
EQUIPMENT: Picture stimuli, touchscreen computer.
TYPES OF LINGUISTIC KNOWLEDGE STUDIED: To date, Principle B of the binding theory (Chapter 4), and control of adjunct clauses (Chapters 4 and 7).
FURTHER READING: Zuckerman et al. 2016; Gerard et al. 2017.

ELICITED PRODUCTION This task involves the researcher prompting the child to respond. For example, the experimenter may begin a sentence and leave the child to complete it, or s/he may ask a question along the lines of 'Tell me what they are doing?' (with reference to a picture or a set of toys).

AGE RANGE FOR WHICH THE TASK IS SUITABLE: 2½ years and up.
EQUIPMENT: Video-/audio-recorder.
TYPES OF LINGUISTIC KNOWLEDGE STUDIED: Wide range of syntactic, semantic and phonological knowledge.
FURTHER READING: Thornton 1996.

Picture Identification The child is asked to choose one of a set of two or more pictures as correct for a stimulus sentence. This task involves minimal effort on the part of the child, as the child has only to point to the picture that s/he thinks is the right one.

Age range for which the task is suitable: 3 years and up.
Equipment: Stimulus pictures; video-/audio-recorder (optional).
Types of linguistic knowledge studied: Wide range of lexical, syntactic and semantic knowledge.
Further Reading: Gerken and Shady 1996.

Priming Tasks Exposure to a word or a syntactic structure 'primes' related words and similar structures. Two types of priming studies have been done. The first looks at priming of syntactic structures (for example the structure NP PP, in a sentence such as *The girl gave book to her friend*). The results have been used to argue for the independence of syntactic representations from specific lexical items for young children (Snedeker and Thothathiri 2008). The second probes syntactic restrictions, such as those imposed by the binding theory. In a sentence such as *His mistress told the dog to play by himself*, the phrase *the dog* is co-referential with *himself*. Hearing the word *dog* activates (primes) semantic and phonological networks of related words. The prediction is that reaction times would be shorter to say a related word (such as *cat*) than an unrelated word (such as *snow*), when it is presented simultaneously with the word *himself*, since *himself* is linked to the word *dog*. This prediction has been borne out with adults, and experiments have been adapted for child participants. For example, a picture of an animal may be used as the associated word in the activated network and the child may be asked to judge whether the picture is of a live thing or not (McKee 1996).

Age range for which the task is suitable: 3 or 4 years and up.
Equipment: A laboratory equipped for audio and visual presentation and for reaction time monitoring.
Types of linguistic knowledge studied: In principle, a wide range of syntactic phenomena, though the actual studies are limited.
Further Reading: McKee 1996; Snedeker and Thothathiri 2008; Marinis 2010; Vasilyeva et al. 2012.

Question Comprehension In a typical question comprehension experiment, the child is asked to answer a question that could be construed in more than one way. For example, given a context story in which an elephant asks a tiger if the tiger could help a horse to do something, the question *Who did the elephant want to help?* has two possible correct answers: *the tiger* and *the horse*. In other cases, only one possible answer is allowed by the grammar, and the choice of that answer is taken as indicating knowledge of the constraint that blocks the alternative answer. For example, if the question *Who did the elephant ask before helping* follows a story in which an elephant asks a tiger if she should help and horse, and then she does help the horse, only one answer is possible in the adult grammar (the answer *the tiger*), because question formation is not possible from within an adverbial clause (see Figure A.1)

*Figure A.1 A typical story for a question response task testing the constraint on extraction from a temporal clause*

*Story:*
(a) The elephant liked to work.
(b) She asked the tiger: 'Should I help the horse carry those heavy boxes?'
(c) The tiger said yes, so the elephant helped the horse.
(d) The elephant was tired at the end of it all.
Question: Who did the elephant ask before helping?

Age range for which the task is suitable: 3 years and up.
Equipment: Video-/audio-recorder (optional, but desirable).
Types of linguistic knowledge studied: This paradigm has been used to study children's awareness of constraints on question formation.
Further Reading: de Villiers and Roeper 1996.

Repetition In this task, the researcher says a word, phrase or sentence and the child repeats it. The data is analysed for the changes that are made between the stimulus and the response, with the goal of identifying those parts of the grammar that are problematic for the child.

Age range for which the task is suitable: 2 years and up.
Equipment: Video-/audio-recorder.
Types of linguistic knowledge studied: The task has mostly been used to probe syntactic and phonological knowledge.
Further Reading: Lust et al. 1996.

SELF-PACED READING/LISTENING The subject reads or listens to a sentence or sentences presented in a word-by-word or phrase-by-phrase manner. The subject initiates each new word/phrase by a button press. The patterns of response time to initiate the next word are interpreted as indicating difficulty (longer reaction times) or relative ease (shorter reaction times). For example, a longer reaction time to initiate the word after *fruit* in the question *What did the girl eat fruit with?* is interpreted as an indication that the subject has misconstrued *what* as object of *eat*, only to discover her/his error when s/he access the true object *fruit*.

AGE RANGE FOR WHICH THE TASK IS SUITABLE: 8 years and up for self-paced reading; 6–7 years and up for self-paced listening.
EQUIPMENT: Computer programmable for the procedure.
TYPES OF LINGUISTIC KNOWLEDGE STUDIED: The task has mostly been used to probe syntactic phenomena.
FURTHER READING: Aaronson and Ferres 1984; Marinis 2010.

TRUTH VALUE JUDGEMENT This task requires that the child judges statements made by a character as 'True' or 'Not true' (or the equivalent), after the child and the character has heard a story. A danger with the task is young children's propensity to respond positively (the Pollyanna bias), when the correct answer is negative.

AGE RANGE FOR WHICH THE TASK IS SUITABLE: 3 years and up.
EQUIPMENT: Doll/toy animals and other props to accompany the story; video-/audio-recorder (optional).
TYPES OF LINGUISTIC KNOWLEDGE STUDIED: The task has been used to probe syntactic and semantic knowledge in areas such as the binding theory and quantification.
FURTHER READING: Gordon 1996; McKersher and Jaswal 2012.

## LOW TASK DEMAND METHODS

HEAD TURN PROCEDURE Head turn experiments have frequently been used to assess phonetic and phonological knowledge. An adult is seated with an infant on his or her lap. The stimuli are presented (non-simultaneously) from two different sides in front of the infant, and the experimenter records when the infant turns her/his head towards a stimulus (see Figure A.2). The cumulative looking time to a given stimulus is reported. Alternatively, the difference in number of turns can be interpreted as presence/absence of the linguistic parameter represented by the stimulus (for example, failure to head turn for vowels within a language's 'magnet' (Chapter 2), can be interpreted as an infant's knowledge of the boundaries of the magnet in that language).

AGE RANGE FOR WHICH THE TASK IS SUITABLE: 4 to 6 months–1 year.
EQUIPMENT: Computer for audio presentation from different sides; soundproof booth in which the adult/infant sits.

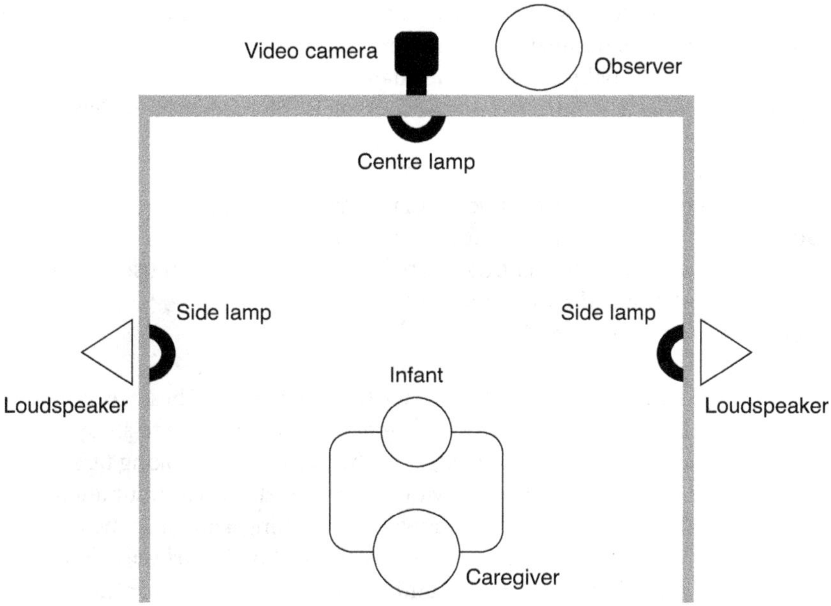


*Figure A.2 The set-up for a preferential looking experiment*
Diagram supplied by Mits Ota.

Types of linguistic knowledge studied: The task has been used to measure phonological distinctions in a variety of languages, and also some syntactic phenomena.
Further Reading: Golinkoff and Hirch-Pasek 2012: 65–66.

Neuroimaging Methods Since the 2000s, the use of methods which measure activity in the brain as speech is input has burgeoned. The techniques comprise:

- Event Related Potential (ERP), which measures the electrical activity in the brain using electrodes positioned on the head.
- Magnetoencephalography (MEG), which detects the magnetic field associated electrical neural activity in the brain.
- Functional Magnetic Resonance Imaging (fMRI) and Functional Near Infrared Spectroscopy (fNIRS) both of which measure changes in blood oxygenation, which correlates with brain activity.

Age range for which the tasks are suitable: All ages.
Equipment: Laboratory suitable for the experiment.
Types of linguistic knowledge studied: All types.
Further Reading: Männel and Friederici 2008; Rispens and Krikhaar 2010; Kovelman 2012.

NON-NUTRITIVE SUCKING A baby is wired to a device that records when the baby sucks on a dummy. The number of sucks over a time period is established for a given stimulus, and when the baby becomes habituated (measured by a fall in sucking rate), the stimulus is changed. A renewed rate of sucking is interpreted as evidence that the infant is aware of the change in stimulus.

AGE RANGE FOR WHICH THE TASK IS SUITABLE: 1–6 months.
EQUIPMENT: Laboratory suitable for the experiment.
TYPES OF LINGUISTIC KNOWLEDGE STUDIED: The task has been used to measure awareness of phonetic distinctions between sounds.
FURTHER READING: Fennell 2012.

PREFERENTIAL LOOKING AND LOOKING WHILE LISTENING These techniques are similar conceptually to the head turn technique, and they are used to study the development of syntactic and lexical knowledge. In the preferential looking task, the child is presented with two screens on which videos are played, with a simultaneous audio stimulus. For example, one video may show a boy washing a girl, and the other video may show a girl washing a boy accompanied by the audio *The girl is washing the boy*. The length of looking time is measured, and longer looking times for the screen that matches the audio can be interpreted as evidence of knowledge of the lexical items or syntactic structures presented. In the looking while listening technique, still pictures are used instead of videos.

AGE RANGE FOR WHICH THE TASK IS SUITABLE: 15 months and up, depending on the focus of the study.
EQUIPMENT: Laboratory suitable for the presentation of video and audio stimuli.
TYPES OF LINGUISTIC KNOWLEDGE STUDIED: Various.
FURTHER READING: Hirsh-Pasek and Golinkoff 1996; Fernald et al. 2008; Piotroski and Naigels 2012.

VISUAL WORLD PARADIGM This technique was developed in the 1990s. The child's eye movements and fixations across a visual array are recorded as they listen to a stimulus.

AGE RANGE FOR WHICH THE TASK IS SUITABLE: 3 years and up.
EQUIPMENT: An eye-tracker; objects in the visual display. A video camera trained on the participant's face may be substituted for the eye-tracker, with subsequent hand-coding of the eye movements.
TYPES OF LINGUISTIC KNOWLEDGE STUDIED: This technique has primarily been used to examine children's analysis of temporary ambiguity; see Chapter 7, p. 121.
FURTHER READING: Sedivy 2010; Trueswell 2012.

## BIBLIOGRAPHY TO APPENDIX 1


Aaronson, D. and S. Ferres (1984), 'The word-by-word reading paradigm: A experimental and theoretical approach', in D. Kieras and M. Just (eds), *New methods in reading comprehension research*, Hillsdale, NJ: Lawrence Erlbaum Associates.

Blom, E. and S. Unsworth (eds) (2010), *Experimental methods in language acquisition research*, Amsterdam, the Netherlands: John Benjamins.

Corrigan, R. (2012), 'Using the CHILDES database', in E. Hoff (ed.), *Research methods in child language*, Chichester: Wiley-Blackwell, pp. 271–284.

Demuth, K. (1996), 'Collecting spontaneous production data', in D. McDaniel, C. McKee and H. S. Cairns (eds), *Methods for assessing children's syntax*, Cambridge, MA: The MIT Press, pp. 3–22.

de Villiers, J. and T. Roeper (1996), 'Questions after stories: On supplying context and eliminating it as a variable', in D. McDaniel, C. McKee and H. S. Cairns (eds), *Methods for assessing children's syntax*, Cambridge, MA: The MIT Press, pp. 163–188.

Fennell, C. (2012), 'Habituation procedures', in E. Hoff (ed.), *Research methods in child language*, Chichester: Wiley-Blackwell, pp. 3–16.

Fernald, A., R. Zangl, A. Portilli and V. Marchman (2008), 'Looking while listening: Using eye movements to monitor spoken language comprehension by infants and young children', in I. Sekerina, E. Fernández and H. Clahsen (eds), *Developmental psycholinguistics: On line methods in children's language processing*, Amsterdam, the Netherlands: John Benjamins, pp. 97–135.

Gerard, J., J. Lidz, S. Zuckerman and M. Pinto (2017), 'Similarity-based interference and the acquisition of adjunct control', *Frontiers in Psychology*, 8, article 1822.

Gerken, L. A. and M. Shady (1996), 'The picture selection task', in D. McDaniel, C. McKee and H. S. Cairns (eds), *Methods for assessing children's syntax*, Cambridge, MA: The MIT Press, pp. 125–146.

Golinkoff, R. and K. Hirsh-Pasek (2012) 'Methods for studying language in infants: Back to the future', in E. Hoff (ed.), *Research methods in child language*, Chichester: Wiley-Blackwell, pp. 60–75.

Goodluck, H. (1996), 'The act-out task', in D. McDaniel, C. McKee and H. S. Cairns (eds), *Methods for assessing children's syntax*, Cambridge, MA: The MIT Press, pp. 147–162.

Gordon, P. (1996), 'The truth-value judgment task', in D. McDaniel, C. McKee and H. S. Cairns (eds), *Methods for assessing children's syntax*, Cambridge, MA: The MIT Press, pp. 211–232.

Hirsh-Pasek, K. and R. Golinkoff (1996), 'The intermodal preferential looking paradigm: A window onto emerging language comprehension', in D. McDaniel, C. McKee and H. S. Cairns (eds), *Methods for assessing children's syntax*, Cambridge, MA: The MIT Press, pp. 105–124.

Hoff, E. (ed.) (2012), *Research methods in child language*, Chichester: Wiley-Blackwell.

Kovelman, I. (2012), 'Neuroimaging methods', in E. Hoff (ed.), *Research methods in child language*, Chichester: Wiley-Blackwell, pp. 43–59.

Lieven, E. and H. Behrens (2012), 'Dense sampling', in E. Hoff (ed.), *Research methods in child language*, Chichester: Wiley-Blackwell, pp. 226–239.

Lust, B., S. Flynn and C. Foley (1996), 'What children know about what they say: Elicited imitation as a research method for assessing children's syntax', in D. McDaniel, C. McKee and H. S. Cairns (eds) (1996), *Methods for assessing children's syntax*, Cambridge, MA: The MIT Press, pp. 55–76.

McDaniel, D., C. McKee and H. S. Cairns (eds) (1996), *Methods for assessing children's syntax*, Cambridge, MA: The MIT Press.

McKee, C. (1996), 'On-line methods', in D. McDaniel, C. McKee and H. S. Cairns (eds), *Methods for assessing children's syntax*, Cambridge, MA: The MIT Press, pp. 189–208.

McKersher, D. and V. Jaswal (2012), 'Using judgment tasks to study language knowledge', in E. Hoff (ed.), *Research methods in child language*, Chichester: Wiley-Blackwell, pp. 149–161.

MacWhinney, B. (2000), *The CHILDES project: Tools for analysing talk*, 3rd edn, Mahwah, NJ: Lawrence Erlbaum Associates.

Männel, C. and A. Friederici (2008), 'Event-related potentials as a window to children's language processing: From syllables to sentences', in I. Sekerina, E. Fernández and H. Clahsen (eds), *Developmental psycholinguistics: On line methods in children's language processing*, Amsterdam, the Netherlands: John Benjamins, pp. 29–72.

Marinis, T. (2010), 'Using on-line methods in language acquisition research', in S. Unsworth and E. Blom (eds), *Experimental methods in language acquisition research*, Amsterdam, the Netherlands: John Benjamins, pp. 139–162.

Naigles, L. (2012), 'Not sampling, getting it all', in E. Hoff (ed.), *Research methods in child language*, Chichester: Wiley-Blackwell, pp. 240–253.

Piotroski, J. and L. Naigles (2012), 'Intermodal preferential looking', in E. Hoff (ed.), *Research methods in child language*, Chichester: Wiley-Blackwell, pp. 17–28.

Rispens, J. and E. Krikhaar (2010), 'Using event-related potentials in language acquisition research', in E. Blom and S. Unsworth (eds), *Experimental methods in language acquisition research*, Amsterdam, the Netherlands: John Benjamins.

Sedivy, J. (2010), 'Using eye-tracking in language acquisition research', in E. Blom and S. Unsworth (eds), *Experimental methods in language acquisition research*, Amsterdam, the Netherlands: John Benjamins, pp. 115–138.

Sekerina, I., E. Fernández and H. Clahsen (eds) (2008), *Developmental psycholinguistics: On-line methods in children's language processing*, Amsterdam, the Netherlands: John Benjamins.

Snedeker, J. and M. Thothathiri (2008), 'What lurks beneath: Syntactic priming during language comprehension', in I. Sekerina, E. Fernández and H. Clahsen (eds), *Developmental psycholinguistics: On-line methods in children's language processing*, Amsterdam, the Netherlands: John Benjamins, pp. 137–167.

Stromswold, K. (1996), 'Analyzing children's spontaneous speech', in D. McDaniel, C. McKee and H. S. Cairns (eds), *Methods for assessing children's syntax*, Cambridge, MA: The MIT Press, pp. 23–54.

Thornton, R. (1996), 'Elicited production', in D. McDaniel, C. McKee and H. S. Cairns (eds), *Methods for assessing children's syntax*, Cambridge, MA: The MIT Press, pp. 77–102.

Trueswell, J. (2012), 'Studying language processing using eye movements', in E. Hoff (ed.), *Research methods in child language*, Chichester: Wiley-Blackwell, pp. 177–190.

Unsworth, S. and E. Blom (eds) (2010), *Experimental methods in language acquisition research*, Amsterdam, the Netherlands: John Benjamins.

Vasilyeva, M., H. Waterfall and L. Gómez (2012), 'Using priming procedures with children', in E. Hoff (ed.), *Research methods in child language*, Chichester: Wiley-Blackwell, pp. 162–176.

Zuckerman, S., M. Pinto, E. Koutamanis and Y. van Sprijk (2016), 'A new method for language comprehension reveals better performance on passive and principle B constructions', in *Proceedings of the Annual 40th Boston University Conference on Language Development*, Boston, MA: Cascadilla Press.

# APPENDIX 2

## RESOURCES FOR CHILD LANGUAGE RESEARCH

The publications listed here reflect the theoretical approach of this book, although many (for example, the journal *Cognition*) also present alternative approaches and debates between opposing viewpoints.

### JOURNALS DEVOTED TO LANGUAGE ACQUISITION/DEVELOPMENT

*First Language*
*Journal of Child Language*
*Language Acquisition*
*Language Learning and Development*

### JOURNALS THAT REGULARLY CARRY ARTICLES ON CHILD LANGUAGE

*Cognition*
*Language*
*Language, Cognition and Neuroscience* (formerly *Language and Cognitive Processes*)
*Lingua*
*Linguistic Inquiry*

### CONFERENCES WITH PUBLISHED PROCEEDINGS

*Annual Boston University Conference on Language Development* (*BUCLD*). Annual.
*Association for the Study of Language Acquisition* (*ASAL*). Triannual.
*Generative Approaches to Language Acquisition* (*GALA*). Biannual.
*Generative Approaches to Language Acquisition North America* (*GALANA*). Biannual.
*Papers and Reports on Child Language* (Stanford University). Annual.
*Tokyo Conference on Psycholinguistics* (*TCP*). Annual.

# BIBLIOGRAPHY


Adam, G. and O. Bat-El (2009), 'When do universal preferences emerge in language development: The acquisition of Hebrew stress', *Brill's Annual of Afroasiatic Languages and Linguistics*, 1, 255–282.

Adger, D. (2003), *Core syntax: A minimalist approach*, Oxford: Oxford University Press.

Aksu-Koç, A. and D. I. Slobin (1985), 'The acquisition of Turkish', in D. I. Slobin (ed.), *The cross-linguistic study of language acquisition*, vol. 1, Hillsdale, NJ: Lawrence Erlbaum Associates, pp. 839–878.

Allen, G. and S. Hawkins (1980), 'Phonological rhythm: Definition and development', in G. Yeni-Komshian, J. Kavanagh and C. Ferguson (eds), *Child Phonology*, vol. 1: *Production*, New York: Academic Press, pp. 227–252.

Ambridge, B. and E. Lieven (2011), *Child language acquisition: Contrasting theoretical approaches*, Cambridge: Cambridge University Press.

Anderson, D. (2005), *The acquisition of tough movement in English*, PhD dissertation, University of Cambridge.

Applegate, J. (1961), 'Phonological rules in a subdialect of English', *Word*, 17, 186–193.

Archibald, J. (ed.) (1995), *Phonological acquisition and phonological theory*, Hillsdale, NJ: Lawrence Erlbaum Associates.

Arnold, J., S. Brown-Schmidt and J. Trueswell (2007), 'Children's use of gender and order of mention during pronoun comprehension', *Language and Cognitive Processes*, 22, 527–565.

Arnold, J., T. Wasow, T. Losongo and R. Ginstrom (2000), 'Heaviness and newness: The effects of structural complexity and discourse status on constituent ordering', *Language*, 76, 28–55.

Aronoff, M., I. Meir, C. Padden and W. Sandler (2008), 'The roots of linguistic organization in a new language', *Interaction Studies*, 9, 133–153.

Baauw, S. (2002), *Grammatical features and the acquisition of reference: A comparative study of Dutch and Spanish*, London: Routledge.

Baker, C. L. (1979), 'Syntactic theory and the projection problem', *Linguistic Inquiry*, 10, 533–581.

Baker, C. L. and J. McCarthy (eds) (1981), *The logical problem of language acquisition*, Cambridge, MA: The MIT Press.

Barner, D. and J. Snedeker (2005), 'Quantity judgment and individuation: Evidence that mass nouns count', *Cognition*, 97, 41–66.

Becker, M. (2014), *The acquisition of syntactic structure*, Cambridge: Cambridge University Press.

Becker, M. and A.-M. Tessier (2011), 'Trajectories of faithfulness in child-specific phonology', *Phonology*, 28, 163–196.

Becker, M., B. Estigarribia and D. Gylfadottír (2012), 'Tough-adjectives are easy to learn', in A. Biller, E. Chung and A. Kimball (eds), *BUCLD 36 Online Proceedings Supplement*, <http://www.bu.edu/bucld/files/2012/07/becker-36.pdf> (last accessed 26 July 2018).

Bellugi, U., S. Marks, A. Bihrle and H. Sabo (1993), 'Dissociation between language and cognitive functions in Williams syndrome', in D. Bishop and K. Mogford (eds), *Language development in exceptional circumstances*, Hillsdale, NJ: Lawrence Erlbaum Associates, pp. 177–189.

Berko, J. (1958), 'The child's learning of English morphology', *Word*, 14, 150–177.

Berwick, R. (1985), *The acquisition of syntactic knowledge*, Cambridge, MA: The MIT Press.

Bever, T. (1970), 'The cognitive basis for linguistic structures', in J. Hayes (ed.), *Cognition and the development of language*, New York: Wiley, pp. 279–352.

Bishop, D. and K. Mogford (eds) (1993), *Language development in exceptional circumstances*, Hillsdale, NJ: Lawrence Erlbaum Associates.

Bloom, P. (1990), 'Subjectless sentences in child language', *Linguistic Inquiry*, 21, 491–504.

Borer, H. and K. Wexler (1987), 'The maturation of syntax', in T. Roeper and E. Williams (eds), *Parameter setting and language acquisition*, Dordrecht, the Netherlands: Reidel, pp. 123–172.

Borer, H. and K. Wexler (1992), 'Bi-unique relations and the maturation of grammatical principles', *Natural Language and Linguistic Theory*, 10, 147–189.

Boser, K., B. Lust, L. Santelmann and J. Whitman (1992), 'The syntax of VP and CP in early German: The strong continuity hypothesis', *Proceedings of NELS*, 22, 51–65.

Bowerman, M. (1973), *Early syntactic development: A cross-linguistic study with special reference to Finnish*, Cambridge: Cambridge University Press.

Bowerman, M. (1982), 'Reorganization processes in lexical and syntactic development', in E. Wanner and L. Gleitman (eds), *Language acquisition: The state of the art*, Cambridge: Cambridge University Press, pp. 319–346.

Bowerman, M. (1987), 'Commentary: Mechanisms of language acquisition', in B. MacWhinney (ed.), *Mechanisms of language acquisition*, Hillsdale, NJ: Lawrence Erlbaum, pp. 444–466.

Bowerman, M. and S. Choi (2001), 'Shaping meanings for language: Universal and language-specific in the acquisition of spatial semantic categories', in M. Bowerman and S. Levinson (eds), *Language acquisition and conceptual development*, Cambridge: Cambridge University Press, pp. 475–511.

Bowerman, M. and W. Croft (2008), 'The acquisition of English causative alternations', in M. Bowerman and P. Brown (eds), *Cross-linguistic perspectives on argument structure: Implications for learnability*, Hillsdale, NJ: Lawrence Erlbaum, pp. 279–307.

Bowerman, M. and S. Levinson (eds) (2001), *Language acquisition and conceptual development*, Cambridge: Cambridge University Press.

Boysson-Bardies, de B., M. Vihman, L. Roug-Hellichius, C. Durand, I. Landberg and F. Arao (1992), 'Material evidence of infant selection from the target language: A cross-linguistic phonetic study', in C. Ferguson, L. Menn and C. Stoel-Gammon (eds), *Phonological development: Models, research implications*, Timonium, MD: York Press, pp. 369–391.

Bracken, H. (1983), *Mind and language*, Dordrecht, the Netherlands: Foris.

Braine, M. (1963), 'The ontogeny of English phrase structure: The first phase', *Language*, 39, 1–13.

Braine, M. (1976), 'Review article: Smith, 1973', *Language*, 52, 489–498.

Braine, M., R. E. Brody, S. M. Fisch, M. J. Weisberger and M. Blum (1990), 'Can children use a verb without exposure to its argument structure?', *Journal of Child Language*, 17: 2, 313–342.

Branigan, H. and K. Messenger (2016), 'Consistent and cumulative effects of syntactic experience in children's sentence production: Evidence for error-based implicit learning', *Cognition*, 157, 250–256.

Brooks, P. and M. Braine (1996), 'What do children know about the universal quantifiers *all* and *each*?', *Cognition*, 60: 3, 235–268.

Brooks, P., M. Braine, G. Jia and M. da Graca Dias (2001), 'Early representations for *all, each* and their counterparts in Mandarin Chinese and Portuguese', in M. Bowerman and S. Levinson (eds), *Language acquisition and conceptual development*, Cambridge: Cambridge University Press, pp. 316–339.

Brown, R. (1973), *A first language*, Cambridge, MA: Harvard University Press.

Brown, R. (1977), 'Introduction', in C. Snow and C. Ferguson (eds), *Talking to children*, Cambridge: Cambridge University Press, pp. 1–27.

Brown, R. and C. Hanlon (1970), 'Derivational complexity and order of acquisition in child speech', in J. Hayes (ed.), *Cognition and the development of language*, New York: Wiley, pp. 11–54.

Cairns, H., D. McDaniel and J. Ryan Hsu (1993), 'A reply to "children are in control" (Sherman and Lust 1993)', *Cognition*, 48: 2, 193–194.

Carlson, G. (1990), 'Intuitions, category and structure: Comments on McDaniel and Cairns', in L. Frazier and J. de Villiers (eds), *Language acquisition and language processing*, Dordrecht, the Netherlands: Kluwer, pp. 327–333.

Carnie, A. (2012), *Syntax, a generative account*, 3rd edn, Oxford: John Wiley-Blackwell.

Chien, Y. C. and K. Wexler (1990), 'Children's knowledge of locality restrictions in binding as evidence for the modularity of syntax and pragmatics', *Language Acquisition*, 1, 225–295.

Cho, S. W. (2009), 'Acquisition of Korean reflexive anaphora', in C. Lee, G. Simpson and Y. Kim (eds), *The handbook of East Asian psycholinguistics*, vol. III, Cambridge: Cambridge University Press, pp. 150–159.

Chomsky, C. (1969), *The acquisition of syntax from 5–10*, Cambridge, MA: The MIT Press.

Chomsky, N. (1957), *Syntactic structures*, The Hague: Mouton and Co.

Chomsky, N. (1959), 'Review of *Verbal behavior* by B. F. Skinner', *Language*, 35, 26–58.

Chomsky, N. (1965), *Aspects of the theory of syntax*, Cambridge, MA: The MIT Press.

Chomsky, N. (1981), *Lectures on government and binding*, Dordrecht, the Netherlands: Foris.

Chomsky, N. (1986), *Knowledge of language: Its nature, origin and use*, New York: Praeger Publishers.

Chomsky, N. (1993), 'A minimalist program for linguistic theory', in. K. Hale and S. J. Keyser (eds), *The view from Building 20*, Cambridge, MA: The MIT Press, pp. 1–52.

Chomsky, N. (2005), 'Three factors in language design', *Linguistic Inquiry*, 36, 1–22.

Clahsen, H. and C. Felser (2006), 'Grammatical processing in language learners', *Applied Psycholinguistics*, 27, 3–42.

Clahsen, H. and M. Penke (1992), 'The acquisition of agreement morphology and its syntactic consequences: New evidence on German Child Language from the Simone-corpus', in J. Meisel (ed.), *The acquisition of verb placement: Functional categories and V2 phenomena in language acquisition*, Dordrecht, the Netherlands: Reidel, pp. 181–223.

Clahsen, H., M. Rothweiler, A. Woest and G. Marcus (1992), 'Regular and irregular inflection in the acquisition of German noun plurals', *Cognition*, 45, 225–255.

Clancy, P. (1985), 'The acquisition of Japanese', in D. I. Slobin (ed.), *The cross-linguistic study of language acquisition*, vols 1, 2, 3, Hillsdale, NJ: Lawrence Erlbaum Associates.

Clark, E. (1973), 'What's in a word? On the child's acquisition of semantics in his first language', in T. Moore (ed.), *Cognitive development and language acquisition*, New York: Academic Press, 65–110.

Clark, E. (1982), 'The young word maker: A case study of innovation in the child's lexicon', in E. Wanner and L. Gleitman (eds), *Language acquisition: The state of the art*, Cambridge: Cambridge University Press, pp. 390–425.

Clark, E. (1987), 'The principle of contrast: A constraint on language acquisition', in B. MacWhinney (ed.), *Mechanisms of language acquisition*, Hillsdale, NJ: Lawrence Erlbaum Associates, pp. 1–33.

Clark, E. (2001), 'Emergent categories in first language acquisition', in M. Bowerman and S. Levinson (eds), *Language acquisition and conceptual development*, Cambridge: Cambridge University Press, pp. 379–405.

Clark, E. and B. Hecht (1982), 'Learning to coin agent and instrument nouns', *Cognition*, 12, 1–24.

Clark, E., B. Hecht and R. Mulford (1986), 'Coining complex compounds in English: Affixes and word order in acquisition', *Linguistics*, 24, 7–29.

Clumeck, H. (1980), 'The acquisition of tone', in G. Yeni-Komshian, J. Kavanagh and C. Ferguson (eds), *Child Phonology*, vol. 1: *Production*, New York: Academic Press, pp. 257–275.

Coates, J. (1983), *The semantics of modal auxiliaries*, London: Longman.

Cohen-Sherman, J. and B. Lust (1993), 'Children are in control', *Cognition*, 46: 1, 1–51.

Conroy, A., E. Takahashi, J. Lidz and C. Philips (2009), 'Equal treatment for all antecedents: How children succeed with Principle B', *Linguistic Inquiry*, 40, 446–486.

Cook, V. and M. Newson (2007), *Chomsky's Universal Grammar*, 3rd edn, Maldon, MA: Blackwell Publishing.

Cowart, W. (1997), *Experimental syntax: Applying objective methods to sentence judgments*, Thousand Oaks, CA: Sage Publications.

Crago, M. and S. Allen (2001), 'Early finiteness in Inuktitut: The role of language structure and input', *Language Acquisition*, 9, 59–111.

Crain, S. and C. McKee (1985), 'The acquisition of structural conditions on anaphora', in *Proceedings of the 15th annual meeting of the North East Linguistics Society*, 94–110, Amherst, MA: University of Massachusetts, Graduate Linguistic Students Association.

Crain, S., R. Thornton, C. Boster, L. Conway, D. Lillo-Martin and E. Woodams (1996), 'Quantification without qualification', *Language Acquisition*, 5, 83–153.

Cromer, R. (1970), 'Children are nice to understand: Surface structure clues for the recovery of a deep structure', *British Journal of Psychology*, 61, 367–408.

Cromer, R. (1987), 'Language growth with experience without feedback', *Journal of Psycholinguistic Research*, 16, 223–232.

Curtiss, S. (1977), *Genie: A psycholinguistic study of a modern 'wild child'*, New York: Academic Press.

Davis, K. (1947), 'Final note on a case of extreme social isolation', *American Journal of Sociology*, 52: 5, 432–437.

Deen, K. Ud and N. Timyan (2018), 'Condition C in adult and child Thai', *Language*, 94, 157–190.

DeGraff, M. (ed.) (1999), *Comparative grammar change: The intersection of language acquisition, Creole genesis and diachronic syntax*, Cambridge, MA: The MIT Press.

DeGraff, M. (1999), 'Creolization, language change and language acquisition: An epilogue', in M. DeGraff (ed.), *Comparative grammar change: The intersection of language acquisition, Creole genesis and diachronic syntax*, Cambridge, MA: The MIT Press, pp. 473–543.

Dell, G. (1995), 'Speaking and misspeaking', in L. Gleitman and M. Liberman (eds), *An invitation to cognitive science*, vol. 1: *Language*, Cambridge, MA: The MIT Press, pp. 183–208.

Demetras, M., K. Nolan Post and C. Snow (1986), 'Feedback to first language learners: The role of repetitions and clarification questions', *Journal of Child Language*, 13, 275–292.

Demuth, K. (1989), 'Maturation and the acquisition of the Sesotho passive', *Language*, 65, 56–80.

Demuth, K. (1995), 'The acquisition of tonal systems', in J. Archibald (ed.), *Phonological acquisition and phonological theory*, Hillsdale, NJ: Lawrence Erlbaum Associates, pp. 111–134.

Demuth, K. (1996), 'The prosodic structure of early words', in J. Morgan and K. Demuth (eds), *Signal to syntax*, Mahwah, NJ: Lawrence Erlbaum Associates, pp. 171–184.

Demuth, K., F. Moloi and M. Machobane (2010), 'Three-year-olds' comprehension, production and generalization of Sesotho passives', *Cognition*, 115, 238–251.

Deutch, W., C. Koster and J. Koster (1986), 'What can we learn from children's errors in understanding anaphora?', *Linguistics*, 24, 203–205.

de Villiers, J. (2007), 'The interface of language and Theory of Mind', *Lingua*, 117, 1858–1878.

de Villiers, J. and P. de Villiers (2014), 'The role of language in Theory of Mind development', *Topics in Language Disorders*, 34, 313–328.

de Villiers, J. and T. Roeper (1995), 'Relative clauses are barriers to wh-movement in young children', *Journal of Child Language*, 22, 389–404.

de Villiers, J. and T. Roeper (1996), 'Questions after stories: Supplying context and removing it as a variable', in D. McDaniel, C. McKee and H. Smith Cairns (eds), *Methods for assessing children's syntax*, Cambridge, MA: The MIT Press, pp. 163–188.

de Villiers, J., T. Roeper and A. Vainikka (1990), 'The acquisition of long-distance rules', in L. Frazier and J. de Villiers (eds), *Language acquisition and language processing*, Dordrecht, the Netherlands: Kluwer, pp. 257–297.

de Villiers, J., T. Roeper, L. Bland-Stewart and B. Pearson (2008), 'Answering hard questions: *Wh*-movement across dialects and disorder', *Applied Psycholinguistics*, 29, 67–103.

Donaldson, M. (1978), *Children's minds*, New York: W. W. Norton and Co.

Dresher, B. E. (1999), 'Child phonology, learnability and phonological theory', in W. Ritchie and T. Bhatia (eds), *Handbook of Child Language Acquisition*, San Diego, CA: Academic Press, pp. 299–346.

Drozd, K. (1995), 'Child English pre-sentential negation as metalinguistic exclamatory sentence negation', *Journal of Child Language*, 22, 583–610.

Drozd, K. (2001), 'Children's weak interpretations of universally quantified questions', in M. Bowerman and S. Levinson (eds), *Language acquisition and conceptual development*, Cambridge, UK: Cambridge University Press, pp. 340–379.

Drozd, K. (2002), 'Negative DPs and elliptical negation in child English', *Language Acquisition*, 10, 77–122.

Eilers, R., W. Gavin and W. Wilson (1979), 'Linguistic experience and phonemic perception in infancy: A cross linguistic study', *Child Development*, 50, 14–18.

Eimas, P., E. Siqueland, P. Jusczyk and J. Vigorito (1971), 'Speech perception in infants', *Science*, 171, 303–306.

Eisenberg, S. (2002), 'Interpretation of relative clauses by young children: Another look', *Journal of Child Language*, 29, 177–188.

Elbourne, P. (2005), 'On the acquisition of Principle B', *Linguistic Inquiry*, 36, 333–365.

Elman, J., E. Bates, M. Johnson, A. Karmiloff-Smith, D. Parisi and K. Plunkett (1996), *Rethinking innateness: A connectionist perspective on development*, Cambridge, MA: The MIT Press.

Felix, S. (1987), *Cognition and language growth*, Dordrecht, the Netherlands: Foris.

Felser, C., T. Marinis and H. Clahsen (2003), 'Children's processing of ambiguous sentences: A study of relative clause attachment', *Language Acquisition*, 11, 127–163.

Ferguson, C. and D. Slobin (eds) (1973), *Studies of child language development*, New York: Holt, Rinehart and Winston.

Ferguson, C., L. Menn and C. Stoel-Gammon (eds) (1992), *Phonological development: Models, research implications*, Timonium, MD: York Press.

Fikkert, P. (1994), *On the acquisition of prosodic structure*, Doctoral dissertation, Leiden University, HIL Dissertation Series 6.

Fletcher, P. and M. Garman (eds) (1986), *Language acquisition*, 2nd edn, Cambridge: Cambridge University Press.

Fodor, J. A. (1980), 'On the impossibility of acquiring more powerful structures', in M. Piattelli-Palmarini (ed.), *Language and learning: The debate between Jean Piaget and Noam Chomsky*, Cambridge, MA: Harvard University Press, pp. 142–149.

Fodor, J. D. (1995), 'Comprehending sentence structure', in L. Gleitman and M. Liberman (eds), *An invitation to cognitive science: Language*, vol. 1, 2nd edn, Cambridge, MA: The MIT Press, pp. 209–246.

Fodor, J. D. and W. G. Sakas (2017), 'Learnability', in I. Roberts (ed.), *The Oxford Handbook of Universal Grammar*, Oxford: Oxford University Press, pp. 249–269.

Fortescue, M. (1984/5), 'Learning to speak Greenlandic: A case study of a two-year-old's morphology in a polysynthetic language', *First Language*, 5, 101–113.

Fox, D. and Y. Grodzinsky (1998), 'Children's passive: A view from the *by*-phrase', *Linguistic Inquiry*, 29, 311–332.

Frazier, L. and J. de Villiers (eds) (1990), *Language acquisition and language processing*, Dordrecht, the Netherlands: Kluwer.

Frazier, L. and G. B. Flores d'Arcais (1989), 'Filler-driven parsing: A study of gap filling in Dutch', *Journal of Memory and Language* 28, 331–344.

Frazier, L. and K. Rayner (1982), 'Making and correcting errors during sentence processing', *Cognitive Psychology*, 14, 178–210.

Friedmann, N., A. Belletti and L. Rizzi (2009), 'Relativized relatives: Types of intervention in the acquisition of A-bar dependencies', *Lingua*, 119, 67–88.

Galasso, J. (2001), *The development of English word order: A minimalist approach*, Ms., California State University, Northridge.

Gallego, Á. (2011), 'Parameters', in C. Boeckx (ed.), *The Oxford Handbook of Linguistic Minimalism*, Oxford: Oxford University Press.

Garrett, M. (1980), 'Levels of processing in sentence production', in B. Butterworth (ed.), *Language Production*, vol. 1: *Speech and talk*, London: Academic Press, pp. 177–220.

Gehrke, B. and N. Grillo (2009), 'How to become passive', in K. Grohman (ed.), *Explorations in phase theory: Features, argument and interpretation at the interface*, Berlin: Mouton de Gruyter.

Gelman, S. and J. Byrnes (eds) (1991), *Perspectives on language and thought: Interactions in development*', Cambridge: Cambridge University Press.

Gerard, J., J. Lidz, S. Zuckerman and M. Pinto (2017), 'Similarity-based interference and the acquisition of adjunct control', *Frontiers in Psychology*, 8, article 1822.

Gil, K.-H., H. Marsden and G. Tsoulas (2018), 'Poverty of the stimulus and language acquisition: From ancient philosophy to neuroscience', in C. Wright, T. Piske and M. Young-Scholten (eds), *Mind matters in SLA*, Bristol: Multilingual Matters.

Gleitman, L. (1990), 'Structural sources of verb learning', *Language Acquisition*, 1, 1–63.

Gleitman, L. and B. Landau (eds) (1994), *The acquisition of the lexicon*, Cambridge, MA: The MIT Press and Amsterdam, the Netherlands: Elsevier Science.

Gleitman, L. and M. Liberman (eds) (1995), *An invitation to cognitive science: Language*, vol. 1, 2nd edn, Cambridge, MA: The MIT Press.

Gleitman, L. and E. Newport (1995), 'The invention of language by children: Environmental and biological influences on the acquisition of language', in L. Gleitman and M. Liberman (eds), *An invitation to cognitive science: Language*, 2nd edn, Cambridge, MA: The MIT Press, pp. 1–24.

Gnanadesikan, A. (2004), 'Markedness and faithfulness constraints in child phonology', in R. Kager, J. Pater and W. Zonneveld (eds), *Fixing priorities: Constraints in phonological acquisition*, Cambridge: Cambridge University Press, pp. 73–108.

Goad, H. (1997), 'Consonant harmony in child language: an optimality-theoretic account', in S. J. Hannahs and M. Young-Scholten (eds), *Focus on phonological acquisition*, Amsterdam, the Netherlands: John Benjamins Publishers, pp. 113–142.

Gold, E. (1967), 'Language identification in the limit', *Information and Control*, 10, 447–474.

Goodluck, H. (1981), 'Children's grammar of complement subject interpretation', in S. Tavakolian (ed.), *Language acquisition and linguistic theory*, Cambridge, MA: The MIT Press.

Goodluck, H. (1990), 'Knowledge integration in processing and acquisition: Comments on Grimshaw and Rosen', in L. Frazier and J. de Villiers (eds), *Language acquisition and language processing*, Dordrecht, the Netherlands: Kluwer.

Goodluck, H. (1997), 'Relative clauses in the speech of Adam', *Proceedings of the bi-annual Generative Approaches to Language Acquisition conference*, 51–56.

Goodluck, H. (2001), 'The nominal analysis of children's interpretations of adjunct PRO clauses', *Language*, 77, 494–509.

Goodluck, H. (2002), 'Islands, parsing and learnability: A commentary on some experimental assessments of children's knowledge of island constraints', *Belfast Working Papers in Language and Linguistics*, 15, 22–37.

Goodluck, H. (2007), 'Formal and computational constraints on language development', in E. Hoff and M. Schatz (eds), *The Blackwell Handbook of Language Development*, Malden, MA: Blackwell Publishing Company.

Goodluck, H. (2010), 'Object extraction is not subject to Child Relativized Minimality', *Lingua*, 120, 1516–1521.

Goodluck, H. and D. Behne (1992), 'Development in control and extraction', in J. Weissenborn, H. Goodluck and T. Roeper (eds), *Theoretical issues in language acquisition*, Hillsdale, NJ: Lawrence Erlbaum, pp. 151–171.

Goodluck, H. and N. Kazanina (2020), 'Fragments along the way: Minimalism as an account of stages in first language acquisition', in A. Benítez-Burraco, K. Fujita, K. Hoshi and L. Progovac (eds), *Frontiers in Psychology, issue of The Biology of Language under a Minimalist Lens: Promises, Achievements, and Limits.*

Goodluck, H. and M. Rochemont (eds) (1992), *Island constraints: Theory, acquisition and processing*, Dordrecht, the Netherlands: Kluwer.

Goodluck, H. and L. Solan (2001), 'Un effet du principe C chez l'enfant francophone', *Canadian Journal of Linguistics/Revue linguistique du Canada*, 45: 1–2, 49–62.

Goodluck, H. and D. Stojanović (1996), 'The structure and acquisition of relative clauses in Serbo-Croatian', *Language Acquisition*, 5: 4, 285–315.

Goodluck, H. and S. Tavakolian (1982), 'Competence and processing in children's grammar of relative clauses', *Cognition*, 11, 1–27.

Goodluck, H., A. Eriks-Brophy and D. Stojanović, *Syntactic ability in high-functioning individuals with Down syndrome*, Ms., University of York, University of Toronto and Government of Canada.

Goodluck, H., M. Finney and J. Sedivy (1991), 'Sentence completeness and filler-gap dependency parsing', in P. Coopmans, B. Schouten and W. Zonneveld (eds), *The OTS*

*Yearbook 1991*, Utrecht, the Netherlands: Research Institute for Language and Speech, pp. 19–31.

Goodluck, H., M. Foley and J. Sedivy (1992), 'Adjunct islands and acquisition', in H. Goodluck and M. Rochemont (eds), *Island constraints: Theory, acquisition and processing*, Dordrecht, the Netherlands: Kluwer, pp. 181–194.

Goodluck, H., E. Guilfoyle and S. Harrington (2006), 'Merge and binding in child relative clauses: The case of Irish', *Journal of Linguistics*, 42, 629–661.

Goodluck, H., K. Saah and D. Stojanović (1995), 'On the default mechanism for interrogative binding', *Canadian Journal of Linguistics/La review Canadienne de linguistique*, 40: 4, 377–404.

Goodluck, H., A. Terzi and G. Chocano Diaz (2001), 'The acquisition of control cross-linguistically: Structural and lexical factors in learning to licence PRO', *Journal of Child Language*, 28, 153–172.

Goodluck, H., F. Tsiwah and K. Saah (2017), 'Island constraints are not the result of sentence processing', *Proceedings of the Linguistic Society of America 2017 Annual Meeting.*

Gordon, P. (1985a), 'Evaluating the semantic categories hypothesis: The case of the mass/count distinction', *Cognition*, 20, 209–242.

Gordon, P. (1985b), 'Level ordering in lexical development', *Cognition*, 21, 73–93.

Gordon, P. (1996), 'The truth-value judgment task', in D. McDaniel, C. McKee and H. Cairns (eds), *Methods for assessing children's syntax*, Cambridge, MA: The MIT Press, pp. 211–232.

Gow, D. (2012), 'The cortical organization of lexical knowledge: A dual lexicon model of spoken language processing', *Brain and Language*, 121, 273–288.

Gow, D. and P. C. Gordon (1995), 'Lexical and prelexical influences on word segmentation', *Journal of Experimental Psychology: Human Perception and Performance*, 21, 344–359.

Gow, D., J. Melvold and S. Manuel (n.d.), *How word onsets drive lexical access and segmentation: Evidence from acoustics, phonology and processing*, Ms., Massachusetts General Hospital, Harvard Medical School and Massachusetts Institute of Technology.

Grant, J., V. Valian and A. Karmiloff-Smith (2002), 'A study of relative clauses in Williams syndrome', *Journal of Child Language*, 29, 403–416.

Grégoire, A. (1937), *L'apprentissage du langage. Les deux premières années*, Liège/Paris: Droz.

Grégoire, A. (1947), *L'apprentissage du langage. La troisième année et les années suivantes*, no place of publication in edition: Droz.

Griffiths, P. (1986), 'Early vocabulary', in P. Fletcher and M. Garman (eds), *Language acquisition*, 2nd edn, Cambridge: Cambridge University Press, pp. 279–306.

Grimshaw, J. (1981), 'Form function and the language acquisition device', in C. Baker and J. McCarthy (eds), *The logical problem of language acquisition*, Cambridge, MA: The MIT Press, pp. 165–182.

Grodzinsky, Y. and T. Reinhart (1993), 'The innateness of binding and coreference', *Linguistic Inquiry*, 24, 69–101.

Gropen, J., S. Pinker, M. Hollander, R. Goldberg and R. Wilson (1989), 'The learnability and acquisition of the dative alternation in English', *Language*, 65, 203–257.

Gruber, J. (1967), 'Topicalization in child language', *Foundations of Language*, 3, 37–65.

Gruber, J. (1973), 'Playing with distinctive features in the babbling of infants', in C. Ferguson and D. Slobin (eds), *Studies of child language development*, New York: Holt, Rinehart and Winston, pp. 4–12.

Guasti, M. T. (2016), *Language acquisition: The growth of grammar*, 2nd edn. Cambridge, MA: The MIT Press.

Guasti, M. T. and A. Cardinaletti (2003), 'Relative clause formation in Romance child's production', *Probus*, 15, 47–89.

Guasti, M. T. and U. Shlonsky (1995), 'The acquisition of French relative clauses reconsidered', *Language Acquisition*, 4, 257–276.

Guilfoyle, E. and M. Noonan (1988), 'Functional categories and language acquisition', paper presented at the 13th Annual Boston University Conference on Language Development.

Haegeman, L. and J. Guéron (1999), *English grammar: a generative perspective*, Oxford: Blackwell Publishers.

Hale, K. (1983), 'Warlpiri and the grammar of non-configurational languages', *Natural Language and Linguistic Theory*, 1, 5–47.

Halle, M. and C. Clements. (1983), *Problem book in phonology*, Cambridge, MA: The MIT Press.

Hamburger, H. and S. Crain (1982), 'Relative acquisition', in S. Kuczaj (ed.), *Language development*, vol 1: *Syntax and semantics*, Hillsdale, NJ: Lawrence Erlbaum.

Harley, H. (2011), 'Compounding in Distributed Morphology', in R. Lieber and P. Stekauer (eds), *Oxford Handbook of Compounding*, Oxford: Oxford University Press. DOI: 10.1093/oxfordhb/9780199695720.013.0007.

Harris, T. and K. Wexler (1996), 'The optional-infinite stage in child English: Evidence from negation', in H. Clahsen (ed.), *Generative perspectives on language acquisition: Empirical findings, theoretical considerations and cross-linguistic comparisons*, Amsterdam, the Netherlands: John Benjamins, pp. 1–42.

Hayes, B. (2009), *Introductory phonology*, Malden, MA: Wiley-Blackwell.

Hayes, J. (ed.) (1970), *Cognition and the Development of Language*, New York: Wiley.

Hickman, M. and H. Hendricks (1999), 'Cohesion and anaphora in children's narratives: A comparison of English, French, German and Mandarin Chinese', *Journal of Child Language*, 26, 419–452.

Hirsch, C. and K. Wexler (n.d.), 'By the way, children don't know *by*', Cambridge, MA: Massachusetts Institute of Technology.

Hirsch, C., R. Orfitelli and K. Wexler (2007), 'When *seem* means *think*: The role of the experiencer-phrase in children's comprehension of raising', in A. Belikova, L. Meroni and M. Umeda (eds), *Proceedings of the 2nd conference on generative approaches to language acquisition (GALANA)*, Somerville, MA: Cascadilla Proceedings Project, pp. 135–146.

Hirsh-Pasek, K., D. Kemler Nelson, P. Jusczyk, W. Wright, B. Druss and L. Kennedy (1987), 'Clauses are perceptual units for young infants', *Cognition*, 26, 269–286.

Hochberg, J. (1988), 'Learning Spanish stress', *Language*, 64, 683–706.

Hoekstra, T. and N. Hyams (1998), 'Aspects of root infinitives', *Lingua*, 106, 81–112.

Hornstein, N. and D. Lightfoot (eds) (1981), *Explanation in linguistics: The logical problem of language acquisition*, London, UK: Longman.

Hsu, J., H. Cairns and R. Fiengo (1985), 'The development of grammars underlying children's interpretation of complex sentences', *Cognition*, 20, 25–48.

Hyams, N. (1986), *Language acquisition and the theory of parameters*, Dordrecht, the Netherlands: Reidel.

Hyams, N. (2012), 'Eventivity effects in early grammar: The case of non-finite verbs', *First Language*, 31, 239–269.

Ingram, D. (1986), 'Phonological development', in P. Fletcher and M. Garman (eds), *Language acquisition*, 2nd edn, Cambridge: Cambridge University Press, pp. 223–239.

Inhelder, B. and J. Piaget (1964), *The early growth of logic in the child*, London: Routledge and Kegan Paul.

Ioup, G. (1975), 'Some universals for quantifier scope', *Syntax and Semantics*, 4, 37–58.

Jakobson, R. (1968), *Child language, aphasia and phonological universals*, translated by A. Keiler from *Kindersprache, Aphasie und allgemeine Lautgesetze*, The Hague: Mouton.

Jakobson, R., G. Fant and M. Halle (1952), *Preliminaries to speech analysis*, Technical Report 13, MIT Acoustics Laboratory, (10th printing, Cambridge, MA: The MIT Press, 1972).

Jakubowicz, C. (1984), 'On markedness and binding principles', in C. Jones and P. Sells (eds), *Proceedings of the 14th North Eastern Linguistics Association Meeting*, Amherst, MA: Graduate Linguistics Students Association, University of Massachusetts-Amherst.

Jakubowicz, C. (1994), 'Reflexives in French and Danish: Morphology, syntax and acquisition', in B. Lust, G. Hermon and J. Kornfilt (eds), *Syntactic theory and first language acquisition*, vol. 2, Hillsdale, NJ: Lawrence Erlbaum Associates, pp. 115–144.

Janke, V. (2018a), 'Discourse effects on older children's interpretations of complement control and temporal adjunct control', *Language Acquisition*, 25, 366–391.

Janke, V. (2018b), 'Who is the agent? The influence of pragmatic leads on children's reference assignment in non-obligatory control', *Journal of Child Language*, 45, 442–478.

Johnson, E. (2008), 'Infants use prosodically conditioned acoustic-phonetic cues to extract words from speech', *Journal of the Acoustical Society of America*, 123, EL 144–148.

Jusczyk, P. (1997), *The discovery of spoken language*, Cambridge, MA: The MIT Press.

Jusczyk, P. and J. Bertoncini (1988), 'Viewing the development of speech perception as an innately guided process', *Language and Speech*, 31, 217–238.

Jusczyk, P., A. Cutler and N. J. Redanz (1993), 'Infants' preference for the predominant stress patterns of English words', *Child Development*, 64, 675–687.

Kager, R., J. Pater and W. Zonneveld (eds) (2004), *Fixing priorities: Constraints in phonological acquisition*, Cambridge: Cambridge University Press.

Karmiloff-Smith, A. (1980), 'Psychological processes underlying pronominalization and non-pronominalization in children's connected discourse', in J. Keimanand and A. Ojedo (eds), *Papers from the parasession on pronouns and anaphora*, Chicago, IL: Chicago Linguistic Circle, pp. 231–250.

Karmiloff-Smith, A. (1985), 'Language and cognitive processes from a developmental perspective', *Language and Cognitive Processes*, 1, 61–85.

Kazanina, N. and C. Phillips (2001), 'Coreference in child Russian: Distinguishing syntactic and discourse constraints', in *Proceedings of the 25th Annual Boston University Conference on Language Development*, Somerville, MA: Cascadilla Press, pp. 413–424.

Kazanina, N. and C. Phillips (2007), 'A developmental perspective on the imperfective paradox', *Cognition*, 105, 65–102.

Kearns, K. (2011), *Semantics*, Basingstoke: Palgrave.

Kegl, J., A. Senghas and M. Coppola (1999), 'Creation through contact: Sign language emergence and sign language change in Nicaragua', in M. DeGraff (ed.), *Comparative grammar change: The intersection of language acquisition, Creole genesis and diachronic syntax*, Cambridge, MA: The MIT Press, pp. 179–237.

Kehoe, M. (2000), 'Truncation without shape constraints: The latter stages of prosodic acquisition', *Language Acquisition*, 8, 23–67.

Kennedy, R. (2016), *Phonology: A coursebook*, Cambridge: Cambridge University Press.

Kidd, E., S. Brandt, E. Lieven and M. Tomasello (2007), 'Object relatives made easy: A cross-linguistic comparison of constraints influencing young children's processing of relative clauses', *Language and Cognitive Processes*, 22, 860–897.

Kiparsky, P. (1983), 'From cyclic phonology to lexical phonology', in H. van der Hulst and N. Smith (eds), *The structure of phonological representations*, part 1, Dordrecht, the Netherlands: Foris, pp. 131–175.

Klima, E. and U. Bellugi (1973), 'Syntactic regularities in the speech of children', in C. Ferguson and D. Slobin (eds), *Studies of child language development*, New York: Holt, Rinehart and Winston, pp. 333–353.

Korsah, S. and A. Murphy (2019), 'Tonal reflexes of movement in Asante Twi', *Natural Language and Linguistic Theory*, September, 1–59.

Kučera, H. and W. N. Francis (1967), *Computational analysis of present-day American English*, Providence, RI: Brown University Press.

Kuhl, P. K. and J. D. Miller (1975), 'Speech perception by the chinchilla: Voiced-voiceless distinction in alveolar plosive consonants', *Science*, 190: 4209, 69–72.

Kuhl, P., K. Williams, F. Lacerda, K. Stevens and B. Lindblom (1992), 'Linguistic experience alters phonetic perception in infants by 6 months of age', *Science*, 255, 606–608.

Labelle, M. (1990), 'Predication, wh-movement and the development of relative clauses', *Language Acquisition*, 1, 95–119.

Labelle, M. (1996), 'The acquisition of relative clauses: Movement or no movement?', *Language Acquisition*, 5, 65–82.

Landau, B. and L. Gleitman (1985), *Language and experience*, Cambridge, MA: Harvard University Press.

Landau, I. and R. Thornton (2011), 'Early child control', *Lingua*, 121: 5, 920–941.

Lasky, R., A. Syrdal-Lasky and R. Klein (1975), 'VOT discrimination by four to six and a half month infants from Spanish environments', *Journal of Experimental Child Psychology*, 20, 215–225.

Lasnik, H. and J. Lidz (2017), 'The argument from poverty of the stimulus', in I. Roberts (ed.), *The Oxford Handbook of Universal Grammar*, Oxford: Oxford University Press, pp. 221–248.

Lasnik, H. and J. Uriagereka (2002), 'On the poverty of the challenge', *The Linguistic Review*, 19, 147–150.

Lee, H. and K. Wexler (1987), 'The acquisition of reflexives and pronouns in Korean', paper presented at the 12th Annual Boston University Conference on Language Development.

Legate, J. (2003), 'The configurational structure of a nonconfigurational language', in P. Pica (ed.), *Linguistic Variation Yearbook*, vol. 1:1, 2001, Amsterdam, the Netherlands: John Benjamins, pp. 61–104.

Legate, J. and C. Yang (2002), 'Empirical re-assessment of stimulus poverty arguments', *The Linguistic Review*, 19, 151–162.

Lenneberg, E. (1967), *Biological foundations of language*, New York: Wiley.

Levelt, C. (1994), *On the acquisition of place*, Doctoral dissertation, Leiden University, HIL Dissertation Series 8.

Levelt, C. and M. van Oostendorp (2007), 'Feature co-occurrence constraints in L1 acquisition', in B. Los and M. van Koppen (eds), *Linguistics in the Netherlands 2007*, Amsterdam: Benjamins, pp. 162–172.

Levelt, C., N. Schiller and W. Levelt (1999), 'A developmental grammar for syllable structure in the production of child language', *Brain and Language*, 68, 291–299.

Levy, Y. and A. Vainikka (2000), 'The development of a mixed null subject system: A cross-linguistic perspective with data from the acquisition of Hebrew', *Language Acquisition*, 8, 363–384.

Lew-Williams, C. and J. Saffran (2011), 'All words are not created equal: Expectations about word length guide infant statistical learning', *Cognition*, 122, 241–246.

Li, C. and S. Thompson (1977), 'The acquisition of tone in Mandarin-speaking children', *Journal of Child Language*, 4, 185–199.

Lidz, J. and J. Musolino (2002), 'Children's command of quantification', *Cognition*, 84, 115–154.

Lidz, J., H. Gleitman and L. Gleitman (2003), 'Understanding how input matters: Verb learning and the footprint of universal grammar', *Cognition*, 87, 151–178.

Lieber, R. (2015), *Introducing morphology*, Cambridge: Cambridge University Press.

Lieber, R. and P. Stekauer (eds) (2008), *The Oxford Handbook of Compounding*, Oxford: Oxford University Press.

Lightfoot, D. (1989), 'The child's trigger experience: Degree-0 learnability', *Behavioral and Brain Sciences*, 12, 321–334.

Lightfoot, D. (2017), 'Discovering new variable properties without parameters', *Linguistic Analysis*, 41.

Limber, J. (1973), 'The genesis of complex sentences', in T. Moore (ed.), *Cognitive development and the acquisition of language*, New York: Academic Press, pp. 170–185.

Limber, J. (1976), 'Unravelling competence, performance and pragmatics in the speech of young children', *Journal of Child Language*, 3, 309–318.

Lisker, L. and A. Abrahamson (1970), 'The voicing dimensions: Some experiments in comparative phonetics', in *Proceedings of the sixth international congress of phonetic sciences*, Prague: Academica.

Locke, J. (1983), *Phonological acquisition and change*, New York: Academic Press.

Lust, B., G. Hermon and J. Kornfilt (eds) (1994), *Syntactic theory and first language acquisition*, vol. 2, Hillsdale, NJ: Lawrence Erlbaum Associates.

Lust, B., K. Loveland and R. Kornet (1980), 'The development of anaphora in first language: Syntactic and pragmatic constraints', *Linguistic Analysis*, 6, 359–392.

Lust, B., L. Solan, S. Flynn, C. Cross and E. Schuetz (1986), 'A comparison of null and pronominal anaphora in first-language acquisition', in B. Lust (ed.), *Studies in the acquisition of anaphora*, vol. 1, Dordrecht, the Netherlands: Reidel, pp. 245–277.

Lyons, J. (1977), *Semantics*, Cambridge: Cambridge University Press.

Lyons, J. and R. Wales (eds) (1966), *Psycholinguistic Papers*, Edinburgh: Edinburgh University Press.

McDaniel, D., C. McKee and M. Garrett (2010), 'Children's sentence planning: Syntactic correlates of fluency variations', *Journal of Child Language*, 37, 59–94.

McDaniel, D., C. McKee and H. Smith Cairns (1996), *Methods for assessing children's syntax*, Cambridge, MA: The MIT Press.

McDaniel, D., H. Smith Cairns and J. Ryan Hsu (1990/1991), 'Control principles in the grammar of young children', *Language Acquisition*, 1, 297–335.

McKee, C. (1992), 'A comparison of pronouns and anaphors in Italian and English acquisition', *Language Acquisition*, 2: 1, 21–54.

Macken, M. (1992), 'Where's phonology?', in C. Ferguson, L. Menn and C. Stoel-Gammon (eds), *Phonological development: Models, research implications*, Timonium, MD: York Press, pp. 249–269.

Macnamara, J. (1982), *Words for things*, Cambridge, MA: The MIT Press.

MacWhinney, B. (ed.) (1987), *Mechanisms of language acquisition*, Hillsdale, NJ: Lawrence Erlbaum Associates.

MacWhinney, B. (2000), *The CHILDES project: Tools for analysing talk*, vol. 1: *Transcription Format and Programs*; vol. 2: *The database*, 3rd edn, Mahwah, NJ: Lawrence Erlbaum Publishers.

Maratsos, M. (1974a), 'Children who get worse at understanding the passive: A replication of Bever', *Journal of Psycholinguistic Research*, 3, 65–74.

Maratsos, M. (1974b), 'How children understand missing complement subjects', *Child Development*, 45, 700–706.

Maratsos, M., D. Fox, J. Becker and M. Chalkley (1985), 'Semantic restrictions on children's passives', *Cognition*, 19, 167–191.

Marchand, H. (1969), *The categories and types of English word formation*, 2nd revised edn, Munich: Beck.

Marcus, G. (1993), 'Negative evidence in language acquisition', *Cognition*, 46, 53–85.

Marcus, G. (2004), *The birth of the mind*, New York: Basic Books.

Marcus, G. (2017), *Innateness, alpha-zero and artificial intelligence*, Ms., New York University.

Marcus, G., S. Pinker, M. Ullman, T. Rosen and F. Xu (1992), 'Overregularization in child language', *Monographs of the Society for Research in Child Development*, 57, 4 (serial number 228).

Markman, E. (1989), *Categorization and naming in children: Problems of induction*, Cambridge, MA: The MIT Press.

Meisel, J. (ed.) (1992), *The acquisition of verb placement: Functional categories and V2 phenomena in language acquisition*, Dordrecht, the Netherlands: Reidel.

Menn, L. (1976), *Pattern, control and contrast in beginning speech: A case study of the development of word form and word function*, PhD dissertation, University of Illinois, Urbana.

Menn, L. (1983), 'Development of articulatory, phonetic and phonological capabilities', in B. Butterworth (ed.), *Language production: Development, writing and other language processes*, vol. II, New York: Academic Press.

Menn, L. (2004), 'Saving the baby: Making sure old data survive new theories', in R. Kager, J. Pater and W. Zonneveld (eds), *Fixing priorities: Constraints in phonological acquisition*, Cambridge: Cambridge University Press, pp. 54–72.

Menn, L. and E. Matthei (1992), 'The "two lexicon" account of child phonology', in C. Ferguson, L. Menn and C. Stoel-Gammon (eds), *Phonological development: Models, research implications*, Timonium, MD: York Press, pp. 211–247.

Mielke, J. (2008), *The emergence of distinctive features*, Oxford: Oxford University Press.

Miller, J. and P. Eimas (eds) (1995), *Speech, language and communication: Handbook of perception and cognition*, 2nd edn, San Diego, CA: Academic Press.

Moore, T. (ed.) (1973), *Cognitive development and language acquisition*, New York: Academic Press.

Morgan, J. and K. Demuth (eds) (1996), *Signal to syntax*, Mahwah, NJ: Lawrence Erlbaum Associates.

Naigles, L. (1990), 'Children use syntax to learn verb meanings', *Journal of Child Language*, 17, 357–374.

Nelson, K. (1973), *Structure and strategy in learning to talk*, Monographs of the Society for Research in Child Development, 149.

Nguyen, E. and L. Pearl (2017), 'Do you really mean it? Linking lexical semantic profiles and age of acquisition for the English passive', in Wm. G. Bennett, Lindsay Hracs, and Dennis Ryan Storoshenko (eds), *Proceedings of the 35th West Coast Conference on Formal Linguistics*, Somerville, MA: Cascadilla Proceedings Project, pp. 288–295.

Noveck, I. (2001), 'When children are more logical than adults: Experimental investigations of scalar implicature', *Cognition*, 78, 165–185.

Omaki, A., I. White, T. Goro, J. Lidz and C. Philips (2014), 'No fear of commitment: Children's incremental interpretations in English and Japanese wh-questions', *Language Learning and Development*, 10, 206–233.

Otsu, Y. (1981), *Towards a theory of syntactic development*, PhD dissertation, Massachusetts Institute of Technology.

Palmer, F. (1979), *Modality and the English verb*, London: Longman.

Papafragou, A. (1997), 'Modality in language development: A reconsideration of the evidence', *UCL Working Papers in Linguistics*, 9, 1–31.

Papafragou, A., K. Cassidy and L. Gleitman (2007), 'When we think about thinking: The acquisition of belief verbs', *Cognition*, 105, 125–165.

Pater, J. (2019), 'Generative linguistics and neural networks at 60: Foundation, friction and fusion', *Language*, 95, 41–74.

Pater, J. and A. Werle (2003), 'Direction of assimilation in child consonant harmony', *Canadian Journal of Linguistics/Revue canadienne de linguistique*, 48, 385–408.

Pater, J., C. Stager and J. Werker (2004), 'The perceptual acquisition of phonological contrasts', *Language*, 80, 384–402.

Pérez-Leroux, A. T., M. Pirvulescu and Y. Roberge (2017), *Direct objects and language acquisition*, Cambridge, UK: Cambridge University Press

Peters, S. (1972), 'The projection problem: How is a grammar to be selected?', in S. Peters (ed.), *Goals of linguistic theory*, Englewood Cliffs, NJ: Prentice Hall.

Philip, W. (1996), 'The event quantification analysis of symmetrical interpretation and denial of implausible felicity', in *Proceedings of the 20th Annual Boston University Conference on Language Development*, pp. 564–575.

Philip, W. and P. Coopmans (1996), 'The role of lexical feature acquisition in the development of pronominal anaphora', in W. Philip and F. Wijnen (eds), *Amsterdam series on child language development*, vol. 5, Amsterdam, the Netherlands: Instituut Algemene Taalwetenschap.

Philip, W. and F. Wijnen (eds) (1996), *Amsterdam series on child language development*, vol. 5, Amsterdam, the Netherlands: Instituut Algemene Taalwetenschap.

Piaget, J. (1980), 'The psychogenesis of knowledge and its epistemological significance', in M. Piattelli-Palmarini (ed.), *Language and learning: The debate between Jean Piaget and Noam Chomsky*, Cambridge, MA: Harvard University Press, pp. 23–34.

Piattelli-Palmarini, M. (ed.) (1980), *Language and learning: The debate between Jean Piaget and Noam Chomsky*, Cambridge, MA: Harvard University Press.

Pinker, S. (1984), *Language learnability and language development*, Cambridge, MA: Harvard University Press.

Pinker, S. (1987), 'The bootstrapping problem in language acquisition', in B. MacWhinney (ed.), *Mechanisms for language acquisition*, Hillsdale, NJ: Lawrence Erlbaum.

Pinker, S. (1989), *Learnability and cognition: The acquisition of argument structure*, Cambridge, MA: The MIT Press.

Pinker, S. (1995), 'Why the child holded the baby rabbits: A case study in language acquisition', in L. Gleitman and M. Liberman (eds), *An invitation to cognitive science: Language*, vol. 1, 2nd edn, Cambridge, MA: The MIT Press, pp. 107–133.

Pinker, S. and M. Ullman (2002), 'The past and future of the past tense', *Trends in Cognitive Science*, 6, 456–463.

Platzack, C. (1990), 'A grammar without functional categories: A syntactic study of early Swedish child language', *Nordic Journal of Linguistics*, 13, 107–126.

Prince, A. and P. Smolensky (2004), *Optimality Theory: Constraint interaction in generative grammar*, Malden, MA: Wiley-Blackwell.

Pullum, G. and B. Scholz (2002), 'Empirical assessment of stimulus poverty arguments', *The Linguistic Review*, 19, 8–50.

Pye, C. (1992), 'The acquisition of K'iche' Maya', in D. I. Slobin (ed.), *The cross-linguistic study of language acquisition*, vol. 3, Hillsdale, NJ: Lawrence Erlbaum Associates, pp. 221–308.

Quine, W. (1960), *Word and object*, Cambridge, MA: The MIT Press.

Radford, A. (1990), *Syntactic theory and the acquisition of English syntax*, Oxford: Blackwell.

Radford, A. (2009), *An introduction to English sentence structure*, Cambridge: Cambridge University Press.

Randall, J. (1992), 'The catapult hypothesis: An approach to unlearning', in J. Weissenborn, H. Goodluck and T. Roeper (eds), *Theoretical issues in language acquisition: Continuity and change in development*, Hillsdale, NJ: Lawrence Erlbaum Associates, pp. 93–138.

Reinhart, T. (1976), *The syntactic domain of anaphora*, PhD dissertation, Massachusetts Institute of Technology.

Reinhart, T. (1983), *Anaphora and semantic interpretation*, London: Croom Helm.

Reinhart, T. and E. Reuland (1993), 'Reflexivity', *Linguistic Inquiry*, 24, 657–720.

Reuland, E. (2001), 'Primitives of binding', *Linguistic Inquiry*, 32, 439–492.

Rice, K. and P. Avery (1995), 'Variability in a deterministic model of language acquisition: A theory of segmental elaboration', in J. Archibald (ed.), *Phonological acquisition and phonological theory*, Hillsdale, NJ: Lawrence Erlbaum Associates, pp. 23–42.

Rice, M. (2017), 'Specific Language Impairment: What do we know?', *Adjacent Open Access*, August, 34–35.

Rice, M., J. Tomblin, L. Hoffman, W. Richman and J. Marquis (2004), 'Grammatical tense deficits in children with SLI and non-specific language impairment: Relationship with non-verbal IQ over time', *Journal of Speech and Hearing Research*, 47, 816–834.

Rizzi, L. (1990), *Relativized Minimality*, Cambridge, MA: The MIT Press.

Rizzi, L. (2008), 'On the grammatical basis of language development: A case study', in G. Cinque and R. Kayne (eds), *The Oxford Handbook of Comparative Syntax*, Oxford: Oxford University Press, pp. 70–109.

Roberts, I. (2018), 'Parameters of clausal negation', Colloquium at the University of Essex, 8 November 2018.

Roeper, T. (2007), *The prism of grammar: How child language illuminates humanism.* Cambridge, MA: MIT Press.

Roeper, T. and J. de Villiers (1994), 'Lexical links in the WH-chain', in B. Lust, G. Hermon and J. Kornfilt (eds), *Syntactic theory and first language acquisition*, vol. 2, Hillsdale, NJ: Lawrence Erlbaum Associates, pp. 357–390.

Roeper, T. and B. Rohrbacher (1994), *True pro-drop in English and the principle of economy of projection*, Ms., University of Massachusetts-Amherst.

Roeper, T. and E. Williams (eds) (1987), *Parameter setting*, Dordrecht, the Netherlands: Reidel.

Rondal, J. (1995), *Exceptional language development in Down Syndrome*, Cambridge: Cambridge University Press.

Rose, Y. and C. Champdoizeau (2008), 'There is no innate trochaic bias: Acoustic evidence in favour of the neutral start hypothesis', in A. Gavarró and M. Freitas (eds), *Language acquisition and development: Proceedings of GALA 2007*, Newcastle: Cambridge Scholars Publishing.

Rosenblatt, F. (1962), *Principles of neurodynamics: Perceptrons and the theory of brain mechanisms*, Washington, DC: Spartan Books.

Ruigendijk, E., S. Baauw, S. Zuckerman, N. Vasić, J. de Langue and S. Avrutin (2011), 'A cross-linguistic study on the interpretation of pronouns and children and agrammatic speakers: Evidence from Dutch, Spanish and Italian', in E. Gibson and N. Pearlmutter (eds), *The processing and acquisition of reference*, Cambridge, MA: The MIT Press, pp. 133–155.

Rumelhart, D. and J. McClelland (1986), 'On learning the past tenses of English verbs', in D. Rumelhart, J. McClelland and the PDP Research group (eds), *Parallel distributed processing: Explorations in the microstructures of cognition*, vol. 2, Cambridge, MA: MIT Press, pp. 216–271.

Saah, K. and H. Goodluck (1995), 'Island effects in parsing and grammar: Evidence from Akan', *The Linguistic Review*, 12, 318–409.

Saffran, J. and E. Thiesson (2003), 'When cues collide: Use of stress and statistical cues to word boundaries by 7-to-9-month-old infants', *Developmental Psychology*, 39, 706–716.

Saffran, J., R. Aslin and E. Newport (1996), 'Statistical learning by 8-month-olds', *Science*, New Series, 274, 1926–1928.

Segoe, M. (1965), 'Verbal development in a mongoloid', *Exceptional Children*, 6, 229–275.

Seidl, A., G. Hollich and P. Jusczyk (2003), 'Early understanding of subject and object questions', *Infancy*, 4, 423–436.

Shatz, M. and S. Wilcox (1991), 'Constraints on the acquisition of English modals', in S. Gelman and J. Byrnes (eds), *Perspectives on language and thought: Interactions in development*', Cambridge: Cambridge University Press, pp. 319–353.

Shaw, P. (1991), 'Consonant harmony systems: The special status of coronal harmony', in C. Paradis and J. F. Prunet (eds), *The special status of coronals*, New York: Academic Press, pp. 125–157.

Sheldon, A. (1974), 'The role of parallel function in the acquisition of relative clauses in English', *Journal of Verbal Learning and Verbal Behavior*, 13, 274–281.

Shi, R., J. Werker and J. Morgan (1999), 'Newborn infants' sensitivity to perceptual cues to lexical and grammatical words', *Cognition* 72, B11–21.

Siddiqi, D. (2006), *Minimise exponence: Economy effects on a model of the morphosyntactic component of the grammar*, PhD dissertation, University of Arizona.

Sinclair, A., H. Sinclair and O. Marcellus (1971), 'Young children's comprehension and production of passive sentences', *Archives de Psychologie*, 41, 1–22.

Slobin, D. I. (ed.) (1985), *The cross-linguistic study of language acquisition*, vols 1, 2, 3, Hillsdale, NJ: Lawrence Erlbaum Associates.

Smith, N. (1973), *The acquisition of phonology: A case study*, Cambridge: Cambridge University Press.

Smith, N. (2010), *Acquiring phonology: A cross-generational study*, Cambridge: Cambridge University Press.

Smith, N. and I.-M. Tsimpli (1995), *The mind of a savant: Language learning and modularity*, Oxford: Basil Blackwell.

Smith, P., G. Baker and A. Groat (1982), 'Spelling and a source of information about children's linguistic knowledge', *British Journal of Psychology*, 73, 339–350.

Snow, C. (1972), 'Mothers' speech to children learning language', *Child Development*, 43, 549–563.

Snow, C. (1986), 'Conversations in children', in P. Fletcher and M. Garman (eds), *Language Acquisition*, 2nd edn, Cambridge: Cambridge University Press, pp. 69–89.

Snow, C. and C. Ferguson (eds) (1977), *Talking to children*, Cambridge: Cambridge University Press.

Snyder, W. (2001), 'On the nature of syntactic variation: Evidence from complex predicates and complex word formation', *Language*, 77, 324–342.

Snyder, W. (2007), *Child language: A parametric approach*, Oxford: Oxford University Press.

Solan, L. (1978), *Anaphora in child language*, Doctoral dissertation, University of Massachusetts-Amherst.

Solan, L. (1983), *Pronominal reference: Child language and the theory of grammar*, Dordrecht, the Netherlands: Reidel.

Speidel, G. (1984), 'The acquisition of linguistic structures and cognitive development', in C. Thew and C. Johnson (eds), *Proceedings of the second international congress for the study of child language*, vol. 2, Lanham, MD: University Press of America.

Sprouse, J. and D. Almeida (2012), 'Assessing the reliability of textbook data in syntax: Adger's Core Syntax', *Journal of Linguistics*, 48, 485–520.

Stark, R. (1980), 'Stages of speech development in the first year of life', in G. Yeni-Komshian, J. Kavanagh and C. Ferguson (eds), *Child Phonology*, vol. 1: *Production*, New York: Academic Press, pp. 73–92.

Stephany, U. (1986), 'Modality', in P. Fletcher and M. Garman (eds), *Language acquisition*, 2nd edn, Cambridge: Cambridge University Press, pp. 375–400.

Stromswold, K. (1995), 'The acquisition of subject and object wh-questions', *Language Acquisition*, 4, 5–48.

Swingley, D. and R. Aslin (2002), 'Lexical neighborhoods and the word-form representations of 14-month-olds', *Psychological Science*, 13, 480–484.

Sybesma, R. (ed.) (2017), *Encyclopedia of Chinese language and linguistics*, vol. 1, Leiden, the Netherlands: Brill.

Tanenhaus, M. and J. Trueswell (1995), 'Sentence comprehension', in J. Miller and P. Eimas (eds), *Speech, language and communication: Handbook of perception and cognition*, 2nd edn, San Diego, CA: Academic Press.

Tavakolian, S. (ed.) (1981), *Language acquisition and linguistic theory*, Cambridge, MA: The MIT Press.

Tavakolian, S. (1981), 'The conjoined clause analysis of relative clauses', in S. Tavakolian (ed.), *Language acquisition and linguistic theory*, Cambridge, MA: The MIT Press.

Tessier, A.-M. (2016), *Phonological acquisition: Child language and constraint based-grammar*, London: Palgrave.

Thrasher, R. (1977), *One way to say more by saying less: A study of so-called subjectless sentences*, Tokyo: Eihosha.

Tomasello, M. (2003), *Constructing a language: A usage-based theory of acquisition*, Cambridge, MA: Harvard University Press.

Treiman, R., C. Clifton Jr., A. Mayer and I. Wurm (2003), 'Language comprehension and production', in A. Healy and R. Proctor (eds), *Comprehensive handbook of psychology, Vol. 4: Experimental psychology*, New York: John Wiley and Sons.

Trueswell, J., I. Sekerina, N. Hill and M. Logrip (1999), 'The kindergarten-path effect: Studying online sentence processing in young children', *Cognition*, 73, 89–134.

Tsay, J. (2017), 'Acquisition of tone, L1', in R. Sybesma (ed.), *Encyclopedia* (sic) *of Chinese language and linguistics*, vol. 1, Leiden, the Netherlands: Brill.

Tsimpli, I., M. Papakonstantinou and A. Nicolopoulou (2014), 'Linguistic structure and narrative quality in child L1 production', in N. Lavidas, T. Alexiou and A. Areti (eds), *Major trends in theoretical and applied linguistics: Selected papers from the 20th ISTAL*, pp. 387–403.

Tyler, L. and W. Marslen-Wilson (1981), 'Language processing in children', *Journal of Verbal Learning and Verbal Behavior*, 20, 401–416.

Valian, V. (1990), 'Syntactic subjects in the early speech of American and Italian children', *Cognition*, 40, 21–81.

Valian, V. (2006), 'Young children's understanding of present and past tense', *Language Learning and Development*, 2, 251–276.

van Hout, A. (2008), 'Acquiring perfectivity and telicity in Dutch, Italian and Polish', *Lingua*, 118, 1740–1765.

Vendler, Z. (1967), *Linguistics and philosophy*, Ithaca, NY: Cornell University Press.

Verrips, M. and J. Weissenborn (1992), 'Verb placement in Early German and French: The independence of finiteness and agreement', in J. Meisel (ed.), *The acquisition of verb placement: Functional categories and V2 phenomena in language acquisition*, Dordrecht, the Netherlands: Reidel, pp. 283–331.

Vihman, M. (1978), 'Consonant harmony: Its scope and function in child language', in J. Greenberg (ed.), *Universals of human language*, Stanford, CA: Stanford University Press, pp. 403–442.

Vihman, M. (1992), 'Early syllables and the construction of phonology', in C. Fergusson, L. Menn and C. Stoel-Gammon (eds), *Phonological development: Models, research, implications*, Timonium, MD: York Press, pp. 393–422.

Vihman, M. (2014), *Phonological development*, 2nd edn, Oxford: Basil Blackwell.

Vinnitskaya, I. and K. Wexler (2001), 'The role of pragmatics in the development of Russian aspect', *First Language*, 21, 143–186.

Wagner, L. (2001), 'Aspectual influences on early tense comprehension', *Journal of Child Language*, 28, 661–681.

Wanner, E. and L. Gleitman (eds) (1982), *Language acquisition: The state of the art*, Cambridge: Cambridge University Press.

Weinberg, A. (1990), 'Child grammars – Radically different or more of the same?: Comments on de Villiers, Roeper and Vainikka', in L. Frazier and J. de Villiers (eds), *Language acquisition and language processing*, Dordrecht, the Netherlands: Kluwer, pp. 299–311.

Weissenborn, J., H. Goodluck and T. Roeper (eds) (1992), *Theoretical issues in language acquisition: Continuity and change in development*, Hillsdale, NJ: Lawrence Erlbaum.

Weissenborn, J., H. Goodluck and T. Roeper (1992), 'Old and new problems in the study of language acquisition', in J. Weissenborn, H. Goodluck and T. Roeper (eds), *Theoretical issues in language acquisition: Continuity and change in development*, Hillsdale, NJ: Lawrence Erlbaum, pp. 1–23.\

Weist, R., M. Atanassova, H. Wysocka and A. Pavlak (1999), 'Spatial and temporal systems in child language and thought: A cross-linguistic study', *First Language*, 19, 267–311.

Weist, R., P. Lyytinen, J. Wysocka and M. Atanassova (1997), 'The interaction of language and thought in children's language acquisition: A cross-linguistic study', *Journal of Child Language*, 24, 81–121.

Werker, J. (1995), 'Exploring developmental changes in cross-language speech perception', in L. Gleitman and M. Liberman (eds), *An invitation to cognitive science: Language*, vol. 1, 2nd edn, Cambridge, MA: The MIT Press, pp. 87–106.

Werker, J., J. Gilbert, K. Humphrey and R. Tees (1981), 'Developmental aspects of cross-linguistics speech comparison', *Child Development*, 54, 349–353.

Wexler, K. (1981), 'Some issues in the theory of learnability', in C. Baker and J. McCarthy (eds), *The logical problem of language acquisition*, Cambridge, MA: The MIT Press.

Wexler, K. (1992), 'Some issues in the growth of control', in R. Larson, S. Iatridou, U. Lahiri and J. Higginbotham (eds), *Control and grammar*, Dordrecht, the Netherlands: Kluwer, pp. 253–295.

Wexler, K. (1998), 'Very early parameter setting and the unique checking constraint: A new explanation of the optional infinitive stage', *Lingua*, 106, 23–79.

Wexler, K. (2004), 'Theory of phasal development: Perfection in child grammar', *MIT Working Papers in Linguistics*, 48, 159–209.

Wexler, K. and P. Culicover (1980), *Formal principles of language acquisition*, Cambridge, MA: The MIT Press.

Wexler, K. and M. R. Manzini (1987), 'Parameters and learnability', in T. Roeper and E. Williams (eds), *Parameter setting*, Dordrecht, the Netherlands: Reidel, pp. 41–76.

Whitney, W. D. (1889), *Sanskrit grammar*, 2nd edn, Cambridge, MA: Harvard University Press.

Wijnen, F., E. Kirkhaar and E. den Os (1994), 'The (non)realization of unstressed syllables in children's utterances: Evidence for a rhythmic constraint', *Journal of Child Language*, 21, 59–83.

Williams, E. (1981), 'Argument structure and morphology', *The Linguistic Review*, 1, 81–114.

Wright, C., T. Piske and M. Young-Scholten (eds) (2018), *Mind matters in SLA*, Bristol: Multilingual Matters.

Yang, C. (2016), *The price of linguistic productivity: How children learn to break the rules of language*, Cambridge, MA: The MIT Press.

Yeni-Komshian, G., J. Kavanagh and C. Ferguson (eds) (1980), *Child Phonology*, vol. 1: *Production*, New York: Academic Press.

# INDEX

active filler strategy, 118
  in children, 122–3
Akan, 65, 123
Aktionsart, 82
Al-Sayyid Bedouin Sign Language, 99
Arnold, J. et al., 124
Aronoff, M. et al., 99
aspect, 82
  aspect first hypothesis, 82–3
  imperfective paradox, 83–4

Babbling, 15
Barner, D. and J. Snedeker, 95
Becker, M., 70–1, 75
Berwick, R., 102
binding theory, 50, 55–63
Bloom, P., 54
Borer, H. and K. Wexler, 64, 107
Bowerman, M., 38
Bowerman, M. and S. Choi, 93
Braine, M., 19
Brooks, P. and M. Braine, 88–91
Brooks, P. et al., 92
Brown, R., 33, 100

c-command, 56
Chien, Y. C. and K. Wexler, 61
child phonologies, 17–24
child processing
  ability to recover from errors, 120–2
  resolution strategies, 124–5
  similarity to adult processing, 119–20
child production
  filled pauses in, 126
  similarity to adult production, 126–7
Chinese, 20
  Mandarin, 92
Chomsky, C., 57, 69
Chomsky, N., 2–3, 49, 50, 55, 71–2
Clahsen, H. et al., 39–40
Clark, E., 37–8, 93
competence, 1–2
compounding, 39–41
configurational languages, 43–5
connectionism, 106
Conroy, A. et al., 61
conservation, 108
consonant harmony
  child, 26–8
consonants, 6–8
continuity, 107
control theory, 60, 67–9, 70
count nouns, 94
Crago, M. and S. Allen, 34
Crain, S. and C. McKee, 57
Creole languages, 99
Curtis, S., 99

Danish
de Villiers, J. et al., 66
de Villiers, J. and P. de Villiers, 111
Deen, K. Ud and A. Timyan, 58
Demuth, K. et al., 64
Down syndrome, 109–10
Drodz, K., 85–6, 91
Dutch, 62–3

Eimas, P. et al., 13
empty categories, 46–8
English, acquisition of
  binding theory, 56–61
  control, 67–9, 70
  dative forms, 112
  derivational morphology, 37–8

*eager* vs *easy* constructions, 70–1
inflectional morphology, 33
quantification, 88–92
question formation, 65–7
tense, 80–2

Fodor, J. A., 105
Fodor, J. D. and W. Sakas, 102
Fortescue, M., 34
Fox, D. and Y. Grodzinsky, 64–5
Frazier, L. and G. Flores d'Arcais, 118
functional categories, 78–80
acquisition of, 80–2

garden path sentences, 117–18
Gehrke, B. and N. Grillo, 65
German, 39–40
Gleitman, L. and E. Newport, 99
Gnanadesikan, A., 25–6
Goodluck, H., 69–70
Goodluck, H. and D. Behne, 69
Goodluck, H. et al., 122–3
Gordon, P., 39, 94–5
Government-Binding Theory, 71–2
Griffiths, P., 93

Hirsch, C. et al., 107
Hoekstra, T. and N. Hyams, 35–6
Hsu, J. et al., 69
Hyams, N., 54

immature grammar
vs processing pressure, 128
innateness
arguments for, 98–100
Inuktitut, 34
Italian, 60–1

Johnson, E., 74
Jakobson, R., 15–16, 25
Jakubowicz, C., 63
Janke, V., 69, 125
Japanese, 122

Kannada, 96
Karmiloff-Smith, A., 126
Kazanina, N. and C. Philips, 83–4
Kegl, J. et al., 99
Kidd, E. et al., 128
Korean, 93
Kuhl, P. et al., 10–11

Landau, I. and R. Thornton, 69
Learnability, 102
Legate, J. and C. Yang, 101
Levy, Y. and A. Vainikka, 54
Lidz, J. and Muselino, 92
Lust, B. et al., 57

McDaniel, D. et al., 126
McKee, C., 61
Maratsos, M., 69
Maratsos, M. et al., 64
Marcus, G., 100
markedness, 16
maturation, 107
Minimalist Program, 71–3
modality, 86–8
morphology
derivational, 36–7: acquisition of, 37–40
inflectional, 32–3: acquisition of, 33–6
motherese, 104
movement, 46–8, 63–7
island constraints, 65, 122–3

Naigles, L., 109
negation, 84–6
neural networks, 106
Nicaraguan Sign Language, 99
Noveck, I., 87–8

Obligatory Contour Principle (OCP), 20
Omaki, Y. et al., 122
Optimality Theory (OT), 25–6
Overextension
of word meanings, 93

Papafragou, A. et al., 109–10
passive, 63–5, 86, 108
Pater, J., 106–7
performance, 1–2
Philip, W., 91–2
phonetic features
use in child language, 16–17
Piaget, J., 108
Pinker, S., 94, 106
Portuguese, 92

poverty of the stimulus, 101
principle of unique function, 103
prosodic bootstrapping, 74
Pullum, G. and S. Sholz, 101

quantification, 88–92
  child knowledge of, 88–92
  quantifier raising, 89–91

Reinhart, T., 56
relative clauses, 115–16, 127
Reuland, E., 62
Rice, K. and P. Avery, 27
Rizzi, L., 55, 127, 128
Roeper, T. and B. Rohrbacher, 54
root infinitives, 35–6
Ruigendijk, E. et al., 62
rule ordering, 28
Russian, 82–4

Saah, K. and H. Goodluck, 123
Saffran, J. et al., 100
Seidl, A. et al., 66
semantic bootstrapping, 74
Shi, R. et al., 74
Sinclair, A., 108
Smith, N., 17–19
Smith N. and I.-M. Tsimpli, 109
Smith, P., 23–4
Snow, C., 104
Snyder, W., 41
Solan, L., 58, 124–5
speech perception, 10–14
stress, 21–4
Stromswold, K., 66
subcategorization, 48–9
  child errors of, 112
subjectless sentences, 54–5
subset principle, 102–3, 123–4
sufficiency principle, 112
syntactic bootstrapping, 110

telegraphic speech, 51
Thai, 58
thematic roles, 49, 124–5
Theory of Mind (TM), 110
Tomasello, M., 105
tone, 19
Tsay, J., 20
Tyler, L. and W. Marslen-Wilson, 119–20

underextension
  of word meanings, 93
underspecification
  of phonetic features, 27
Universal Grammar (UG), 2–3, 49–50, 69, 101–2, 104, 107
Usage-based approaches to learning, 105

Valian, V., 54, 80–2
vowels, 5–6

Wagner, L., 82
Weissenborn, J., 107
Wexler, K. and R. Manzini, 102
Williams syndrome, 108

X-bar theory, 50

Yang, C., 112

CPSIA information can be obtained
at www.ICGtesting.com
Printed in the USA
BVHW061125290620
582560BV00001B/1

9 781474 458160